The Tactile Internet

SCIENCES

Networks and Communications, Field Director – Guy Pujolle

Internet, Subject Head – Stefano Secci

The Tactile Internet

Coordinated by
Tara Ali-Yahiya
Wrya Monnet

iSTE

WILEY

First published 2021 in Great Britain and the United States by ISTE Ltd and John Wiley & Sons, Inc.

ISTE Ltd
27-37 St George's Road
London SW19 4EU
UK

www.iste.co.uk

John Wiley & Sons, Inc.
111 River Street
Hoboken, NJ 07030
USA

www.wiley.com

Library of Congress Control Number: 2021940372

British Library Cataloguing-in-Publication Data
A CIP record for this book is available from the British Library
ISBN 978-1-78945-020-0

Contents

Foreword

Ian F. AKYILDIZ

School of Electrical and Computer Engineering,
Georgia Institute of Technology, Atlanta, USA

I, Ian F. Akyildiz, have been Ken Byers Chair Professor in Telecommunications at the School of Electrical and Computer Engineering at the Georgia Institute of Technology for the past 35 years. I have vast research experience in wireless communications and many research contributions, including IoT and wireless sensor networks. My h-index is 127 and the total number of citations is 122+K according to Google Scholar as of January 2021. Dr. Tara Ali-Yahiya, Associate Professor at the University of Paris Saclay, and Dr. Wrya Monnet, Assistant Professor at the University of Kurdistan Hewlêr, have been active in this research field for many years. They have introduced this book as an initiative to explain the Tactile Internet to a wide audience in a simple and clear manner.

The Tactile Internet is envisaged to change not only the landscape of network communication but also the lifestyle of society socially and economically. The huge number of use cases introduced by this concept would play a major role towards shaping our imagination for delivering not only data but also skills through the Internet from a source to a remote destination.

This book provides an introduction to the Tactile Internet and its case studies, with its impact on the democratization of haptic applications based on the IEEE 1918.1 standard through teleoperations. The case studies are based on cutting-edge technologies that enable the deployment of the Tactile Internet. 5G, and recently 6G, Software-Defined Networking, the different learning techniques in the artificial intelligence domain, edge computing for service proximity, etc. are all factors that will support the successful deployment of the Tactile Internet. This book is a solid contribution to this research area.

May 2021

Preface

Tara ALI-YAHIYA[1] and Wrya MONNET[2]

[1]*Department of Computer Science, University of Paris-Saclay, France*
[2]*Department of Computer Science and Engineering, University of Kurdistan Hewlêr, Erbil, Iraq*

This book attempts to provide an extensive overview on the Tactile Internet paradigm, which is considered to be the focus of interest around which all the cutting-edge technologies are centered. This is due to its wide applications and use cases that would change our lifestyle. This book is purposely written to appeal to a broad audience and to be of value to anyone who is interested in the Tactile Internet. The audience can be in any domain of computer science, communication and networking. The aim of this book is to offer comprehensive coverage of current state-of-the-art theoretical and technological aspects of the Tactile Internet. The presentation starts from basic principles and proceeds smoothly to more advanced topics. The schemes provided are developed and oriented in the context of very actual closed standards, i.e. IEEE 1918.1.

Organization of this book

This book is organized to follow a methodology of writing depending on how the Tactile Internet is tackled based on the level of difficulty that the audience can face while reading. Hence, we preferred to begin this book with a brief introduction of the tactile concept and its relationship with the Tactile Internet paradigm through cyber-physical systems in Chapter 1. Then, we introduce the architecture of the Tactile Internet from a technical point of view relying mainly on the IEEE 1918.1 standard in Chapter 2. Chapter 3 explains how the success of Tactile Internet deployment relies on different communications technologies, the concept of virtualization and the centralization of intelligence in some parts of the paradigm. These are called key enablers, which will decide the level of success of the Tactile Internet and its use cases. Chapter 4 tackles the 6th Generation of wireless network and its role in boosting the Tactile Internet by adding more intelligence at different levels of the paradigm.

In Chapter 5, IoT technology is reviewed and analyzed, emphasizing its architecture and communication protocols to prepare a background for comparison with the Tactile Internet, both the differences and similarities between them. This chapter attempts to clarify whether the Tactile Internet is an evolution of the IoT or a completely different paradigm. The Internet of Everything (IoE) is also introduced. Its components and differences with the IoT are explained. Later, in Chapter 6, a historical review of telerobotic, teleoperation and telepresence is presented before entering into a detailed explanation of the components and different architectures of teleoperation systems. The two-port system analysis is applied to assess the stability and transparency, which are the performance metrics of the system. A model of a discrete architecture of the teleoperation system is given. The use of the Internet as the medium network for the teleoperation system is presented along with a session initiation protocol to establish a teleoperation session. Finally, a use case of a teleoperation system over the Internet, using two commercial components, is presented. The characteristics and transmission of the haptic data over the Internet are presented in Chapter 7. In this chapter, the perception of haptics in robots is explained: the material, shape and pose recognition. A list of sensor devices for haptic information is given along with their working principles. These are used to build haptic interfaces, where some of the commercial haptic interfaces are also listed. Methods of compression of haptic information for better utilization of the transmission bandwidth is also given in this chapter. The transport protocols of the compressed information are then listed with their properties and conveniences for haptic information communication.

Chapter 8 introduces Wireless Networked Robots and then maps their characteristics, scenarios, and traffic types to the Tactile Internet use cases.

Chapter 9 studies the performance of a teleoperation case study that supports the IEEE 1918.1 architecture while taking 5G as a main transport network. The teleoperation case study investigated the quality of service guarantee for the mission-critical application that requires stringent end-to-end delay. Chapter 10 determines how delay plays a big role in the stability of different types of Tactile Internet applications with the ultra-low latency requirement. It discusses the factors that have an impact on the latency and the recent research work on this subject.

May 2021

List of Acronyms

3GPP	Third-Generation Partnership Project
4G	Fourth Generation
5G	Fifth Generation
6G	Sixth Generation
A2G	Air-to-Ground
AC	Actor Critic
ADMUX	Adaptive Multiplexer for Haptic–Audio–Visual Data Communication
ADSL	Asymmetric Digital Subscriber Line
AI	Artificial Intelligence
AirComp	Air Computation
ALPHAN	Application Layer Protocol for Haptic Networking
AMC	Adaptive Modulation and Coding
AP	Access Point
AR	Auto Regressive
ARMA	Auto Regressive Moving Average
ASIC	Application-Specific Integrated Circuit
ATM	Automated Teller Machine
BBUs	Baseband Units
BCI	Brain–Computer Interaction
BS	Base Station
C-RANs	Cloud Radio Access Networks
CA	Carrier Aggregation
CDF	Cumulative Distribution Function
CDMA	Code Division Multiple Access
CFmMIMO	Cell-Free massive Multiple-Input Multiple-Output
CI	Communication Infrastructure
CMOS	Complementary Metal Oxide Semiconductor
CN	Core Network
CNN	Convolutional Neural Network
CoAP	Constrained Application Protocol
CPP	Control Plane Protocol

CPS	Cyber-Physical Systems
CPU	Central Processing Unit
CSI	Channel State Information
CSMA-CA	Carrier-Sense-Multiple-Access with Collision Avoidance
D2D	Device-to-Device
DCF	Distributed Coordination Function
DCT	Discrete Cosine Transform
DDoS	Distributed Denial of Service
DDS	Data Distribution Service
DNN	Deep Neural Network
DoF	Degree of Freedom
DoS	Denial of Service
DPCM	Differential Pulse Code Modulation
DT	Decision Tree
DTLS	Datagram Transport Layer Security
E2E	End-to-End delay
EC	Edge Computing
EH	Energy Harvesting
eMBB	Enhanced Mobile Broadband
eNodeB	Evolved Node Base
ESA	European Space Agency
ESS	Energy Self-Sustainability
ETP	Efficient Transport Protocol
FiWi	Fiber-Wireless
FPGA	Field Programmable Gate Array
FSM	Finite State Machine
Gbps	Gigabit-per-second
GEO	Geostationary Earth Orbit
GS	Ground Station
GSM	Global System for Mobile communications
GTP	GPRS Tunneling Protocol
H2M	Human-to-Machine
HARQ	Hybrid Automatic Repeated Request
HART	Human–Agent–Robot Teamwork
HEO	High Earth Orbit
HoIP	Haptic over IP
HPC	High Performance Computing
HW	Hardware
IAT	Inter-Arrival Time
IC	Integrated Circuit
ICT	Information and Communication Technologies
IDC	International Data Corporation
IDE	Integrated Development Environment

IEEE	Institute of Electrical and Electronics Engineers
IoDs	IoT Devices
IoE	Internet of Everything
IoT	Internet of Things
IP	Internet Protocol
IPG	Inter-Packet Gap
IR	Infrared
IRTP	Interactive Real-Time Protocol
ISOMAP	ISOmetric MAPping
ITP	Interoperable Telesurgical Protocol
IVR	Immersive Virtual Reality
IW	Indication Weights
JND	Just Noticeable Difference
KNN	K-Nearest Neighbors
KPI	Key Performance Indicator
KPPS	K Packets Per Second
LAN	Local Area Network
LEO	Low Earth Orbit
LNAs	Low Noise Amplifiers
LoS	Line of Sight
LSTM	Long Short-Term Memory
LTE	Long-Term Evolution
LTI	Linear Time Invariant
M2M	Machine-to-Machine
MA	Moving Average
MAC	Medium Access Control
MBSs	Macro Base Stations
MC-IoT	Mission-Critical Internet of Things
MDP	Markov Decision Process
MEC	Mobile Edge Computing
MEMS	Microelectromechanical Systems
MEO	Medium Earth Orbit
MIMO	Multiple-Input Multiple-Output
mMTC	massive Machine-Type Communications
mmWave	millimeter-Wave
MQTT	Message Queuing Telemetry Transport
MRB	Multi-aRmed Bandit
MTU	Maximum Transmission Unit
NAS	Non-Access Stratum
NFV	Network Function Virtualizing
NLoS	Non-Line of Sight
NOMA	Non-orthogonal Multiple Access
NR	New Radio

NS2	Network Simulator-2
NS3	Network Simulator-3
OFDM	Orthogonal Frequency Division Multiplexing
OMA	Orthogonal Multiple Access
OMA-DM	Open Mobile Alliance Device Management
PA	Power Amplifiers
PAHCP	Perception-based Adaptive Haptic Communication Protocol
PCA	Principal Component Analysis
PD	Perceptual Dead-band
PDF	Probability Density Function
PDU	Protocol Data Units
PES	Packetized Elementary Streams
PF-PF	Position–Force-Position–Force
PGW	Packet Gateway
PHY	Physical Layers
PID	Proportional–Integral–Derivative
QoE	Quality of Experience
QoL	Quality of Life
QoS	Quality of Service
R-RA	Radio Resource Allocation
RA	Resource Allocation
RAN	Radio Access Network
RB	Resource Blocks
REST	Representational State Transfer
RF	Radio Frequency
RFID	Radio Frequency Identification
RIS	Reconfigurable Intelligent Surface
RL	Reinforcement Learning
RMS	Root-Mean-Square
RNC	Radio Network Controllers
RNN	Recurrent Neural Network
RRHs	Remote Radio Heads
RTCP	Real-Time Control Protocol
RTNP	Real-Time Network Protocol
RTP	Real-Time Protocol
RTP/I	Real-Time Application Level Protocol for Distributed Interactive Media
RTT	Round Trip Time
SCADA	Supervisory Control And Data Acquisition
SCTP	Synchronous Collaboration Transport Protocol
SDN	Software-Defined Networking
SDU	Service Data Unit

SE	Spectral Efficiency
SGW	Service Gateway
SINR	Signal-to-Interference-plus-Noise Ratio
SIP	Session Initiation Protocol
SLA	Service Level Agreement
SMAC	Social, Mobile, Analytics and Cloud
SNR	Signal-to-Noise Ratio
SPI	Service Plugin Interface
SRC	Semiconductor Research Consortium
SSB	Synchronization Signal Block
STRON	Supermedia Transport for teleoperations over Overlay Networks
SVM	Support Vector Machine
SVR	Support Vector Regression
SW	Software
SWIPT	Simultaneous Wireless Information and Power Transfer
TCP	Transmission Control Protocol
TDD	Time Division Duplex
THz	Terahertz
TI	Tactile Internet
TTI	Transmission Time Interval
UAV	Unmanned Aerial Vehicle
UDP	User Datagram Protocol
UE	User Equipment
UM-MIMO	Ultra-Massive Multiple-Input Multiple-Output
URLLC	Ultra-Reliable Low Latency Communication
USV	Unmanned Surface Vehicle
V2X	Vehicle-to-Everything
VLC	Visible Light Communication
VR/AR	Virtual Reality/Augmented Reality
WET	Wireless Energy Transfer
WG	Working Group
WiFi	Integrated Fiber-Wireless
WLAN	Wireless Local Area Network
WNR	Wireless Networked Robots
WPAN	Wireless Personal Area Network
WPT	Wireless Power Transfer
WSN	Wireless Sensor Network
WTI	Wireless Tactile Internet

1

Introduction to Tactile Internet

Tara Ali-Yahiya

Department of Computer Science, University of Paris-Saclay, France

In broad terms, the Tactile Internet (TI) can be referred to as the interaction between humans and cyber-physical systems by dropping off the distance and ensuring a communication of the order of few milliseconds. Consequently, this would give the illusion that the remote system is too close, while ensuring that the interaction is occurring in a smooth manner. The TI envisions an extremely low latency along with high availability, reliability and +security that will not only revolutionize the technology market, but will also have a high impact on the lifestyle of people, society and business in terms of the vertical industry, according to the definition from the International Telecommunication Union (ITU) (ITU 2014).

However, the TI is viewed as a result of a sequence of cumulative cutting-edge technologies that witnessed great success. These technologies may be considered as the foundation upon which the TI is built. These technologies may include, but are not limited to, diver paradigms in mobile and wireless networks, cloud computing, smart computing, Internet of Things (IoT) and robotics. As a result, the TI can be viewed as the fruit of a multi-disciplinary domain, which has contributed to its progress through various engineering approaches and computational methods to make it function properly.

As a matter of fact, the TI's mode of functioning is tightly linked to the human–machine interaction. Here, the human perception of system operation has a significant impact on the TI's overall performance. We start this chapter by describing the relationship between the human perception of things with regard to the TI and then present an overview on the initiation of the TI, and all of the technologies assisted in

The Tactile Internet,
coordinated by Tara Ali-Yahiya and Wrya Monnet.

its emergence. The objective is to introduce the necessary background and context for understating the TI and its main functionalities.

1.1. Human perception and Tactile Internet

Before getting in depth into the TI paradigm, it is essential to understand its impact on bringing changes to society. In fact, the TI introduced a new disruptive type of data, denoted by haptic data. The haptic data is related more to the human perception of objects through its sensory nervous system. In simple words, the TI proposes to transmit haptic data through the Internet, while the technological background behind this paradigm is more sophisticated and may involve multidisciplinary domains to deploy it and make it work properly. It is common that human haptic systems can perceive close objects through tactile and kinesthetic sensing. What if the objects are located in a remote environment? How can humans touch them and know their nature and properties? How can humans move, rotate and change the position of these objects remotely?

To address these issues, it is noteworthy to be able to recognize the type of haptic data from other types of data which are well-known in the world of Internet. Mainly, haptic data can involve kinesthetic and tactile perceptions, as shown in Figure 1.1. The kinesthetic perception refers to the information that is gathered by the mechanoreceptors located within joint capsular tissues, ligaments, tendons, muscles and skin. Once gathered, the feedback about the position, velocity, angular velocity, force and torque will be treated by the human body. Here, the feedback means the interaction of the human body with the kinesthetic perception, without the use of other senses.

If the kinesthetic is technically interpreted within the context of the human system interaction in the TI, then telesurgery through robots and teleoperation are the best examples to demonstrate it, as the contexts involve human-in-the-loop. This is when the communication and control mechanisms are under the control of humans and require feedback from the distant environment to close the global loop. The effect is represented by sending information about the position, force and movement, so that the robot on the other edge gets instructions to react accordingly. In turn, the robot would send feedback to the human on the other side of the world to assure the continuity of the process. Such feedback is called closed-feedback as it needs to be closed through the information control sent by the distant operator to the human. Certainly, the kinesthetic feedback and a combination of the auditory and visionary give a real human perception, in order to control the operator.

As for tactile information, the mechanoreceptors of the human skin sense various physical information from the environment; this is mainly related to the sense of touch by fingers. Here, tactile feedback can be interpreted by the physical response on an object, from the user input. The user input can be pressing, lifting, touching, etc., and the feedback can be denoted by friction, hardness or warmth that can be felt by the human. The killer application of tactile feedback is in virtual reality and augmented reality, which enable users to physically interact with virtual objects and sense the nature of the objects, locally or remotely. Note that the TI should not be confused with tactile data.

As a matter of fact, kinesthetic and tactile data are new to the Internet and their traffic behavior can vary from low rates to a huge amount of data that can be regulated for transmission through compression and codec techniques. The nature of the feedback determines reaction time of the human, and varies from 1 ms, 10 ms, 100 ms depending on how critical the application is and whether the sensory system of a human is being prepared for that reaction or not. The reaction time is different and the reaction itself can be auditory, visual or a sudden muscular movement. The human perception is complex and requires all of the senses (not only haptic), in order to interact with the surrounding objects. This is why the TI should translate all of these senses and feelings through a whole process of compression, coder/decoder, transmission mechanisms and technology of communication through different use cases.

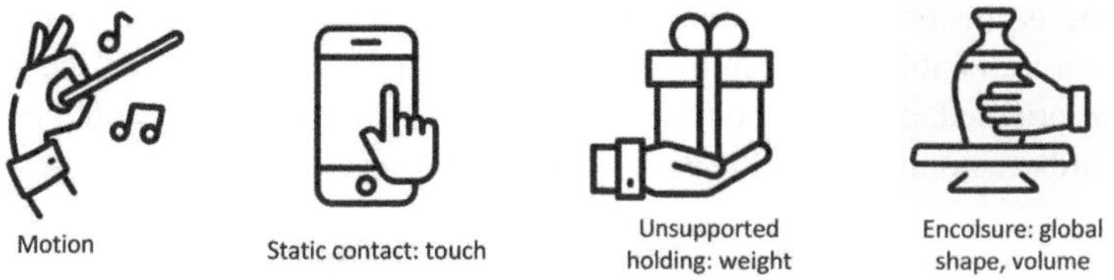

Figure 1.1. *Some haptic perception*

Human sense	Time constant
Muscular interaction	1 ms
Auditory interaction	100 ms
Visual interaction	10 ms
Tactile interaction	1 ms

Table 1.1. *Physiological time constant of different human senses*

1.2. The roadmap towards Tactile Internet

The telecommunication/information and communication technology (ICT) has witnessed fast developments that have paved the way for the TI. The rapid growth of the Internet demonstrated how investment and commitment to research restructured the shape of communication, using fixed infrastructure to connect people, then deploying wireless and mobile infrastructure to have a ubiquitous service anywhere anytime, and connecting different objects to the Internet. Further, the stakeholders implicated in the communication started to vary and take different forms; as mentioned earlier, the communication was restricted to humans only. In addition to human-to-human communication, human-to-machine and machine-to-machine communications came forward.

The technological evolution was beyond the Internet, starting from the time when circuit switching and packet switching appeared. A combination of technological, social and commercial components can be the reason behind the evolution of the infrastructure and services provided to end-users today. The first and/or last mile took different shapes from the perspective of the Internet Service Provider (ISP), bringing the Internet Service to users through the use of the fixed cable-based technology, such as Fiber Optics (FO) and copper telephone lines.

The first or the last mile can currently be offered by wireless and mobile communications, through telecommunication operators. The ubiquitous connectivity is characterized by the employment of mobile and wireless networks, regardless of the technological standardization family, i.e., Institute of Electrical and Electronics Engineers (IEEE) or the 3rd Generation Partnership Project (3GPP). The most recent one is the Fifth Generation (5G) network, which is expected to be the key technology for enabling the deployment of the TI, through its new core network that is service-oriented, depending on the requirement of the users. This would be the first step towards customizing 5G to adapt to the new services introduced by the TI.

As a part of the history of communication, the IoT represented by the machine-to-machine communication is one of the considerable advancements. It encompasses the connectivity of millions of devices to the Internet. The devices can be of any type, but their core mission is to collect data from the environment where they are deployed and send it to the Internet. To process and analyze the generated data, cloud computing through its computing and storage infrastructure for handling this issue would be the best backing paradigm.

The TI can bring all of the technologies of communication and the computing and storage paradigms together to support the transmission of the new haptic data through the Internet. The TI can be considered as the capstone that completes the missing piece of the construction; this is due to the fact that the TI needs to be supported by the efficient technology of communication.

Despite the big transition that occurred in communication technology, i.e. from wired to wireless, the Internet has changed our interaction with the world. For example, from the point of view of User Centered Design (UCD), all of the efforts were spent providing the services of the Internet, while taking user experience into consideration. In the early days, the interaction with information was expressed through the rapidity of retrieving it, the easiness of exchanging it and the tools for searching it and creating it, through the collaborative work of spatially distant actors. While dealing with objects in the IoT, there is a dematerialization of the physical world. Sensors are used to measure the attributes of the real world and actuators are used to collect them. This requires a transfer of data over a network without human-to-human or human-to-machine interaction.

With the newly proposed TI, the human can interact with the machine through the Internet with a return of experience. The major difference between the IoT and the TI is that in the TI, the human is at the center of control; this is why its physiology, represented by the sensor systems of the body, and its psychology, represented by how the user perceives its experience, are crucial. All of the services provided by the TI should ensure that the user has a high Quality of Experience (QoE), so that the interaction between human and machine can be as smooth as possible.

1.3. What is Tactile Internet?

The progress in the development of the access networks, core networks and backbone networks did not happen without the progress of the type of traffic and services provided by the Internet. This may not only require physical changes in the network, but also the introduction of new functionalities and mechanisms that ensure the delivery of the service in an optimal way. If we consider the current traffic recognized by the Internet, we can classify them into video, audio and data. Regardless of the diverse types of applications and services that have appeared over time, the type of traffic did not evolve that much. The introduction of haptic data to the Internet, through the different applications, resulted in innovating the mechanisms dealing with this type of data, in order to transport it through the Internet without any difficulty, as the Internet is not ready to process haptic data yet. Thus, the TI is defined by the IEEE 1918.1 as "a network, or network of networks, for remotely accessing, perceiving, manipulating or controlling real or virtual objects or processes in perceived real time, by humans or machines" is expected to shift the paradigm of the Internet from content delivery to skill delivery (Aijaz *et al.* 2018).

Recently, the IEEE P1918.1 working group started to define the framework, application scenarios and technical concerns. Specifically, the new applications supported by the TI are Industry X.0, Automotive, E-Healthcare, which include a wide range of use cases that may require stringent Quality of Service (QoS), depending on whether they are time critical or not. A simplified architecture of the TI can be shown in Figure 1.2, where three important parts can be identified: the master domain, which is responsible for generating tactile and kinesthetic data; the network domain, which transports the haptic data through the different networks using packets, specifically the access network, core network and the Internet; and finally, the slave domain, where the data will be received and may be processed.

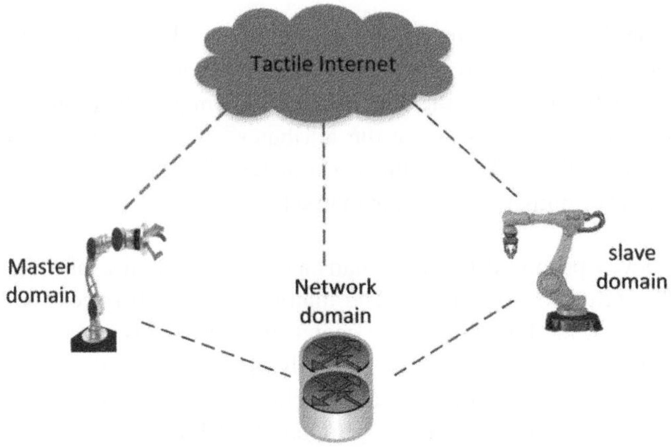

Figure 1.2. *Tactile Internet architecture. For a color version of this figure, see www.iste.co.uk/ali-yahiya/tactile.zip*

The TI is explicitly based on the local action in the master domain and the distant reaction in the slave domain, and vice versa. For this purpose, the TI can be keenly combined with the IoT, as sensors can be deployed in both domains to sense and actuate data (Fettweis 2014). We can imagine a robot in the slave domain that palpates a patient through the touch of a remote doctor in the master domain; the palpitation of the robot is controlled by sensorial gloves worn by the doctor to feel the body of the patient.

Haptic communication is an essential foundation of the TI. As the haptic devices permit the users to feel, touch and manipulate things over real and virtual fields, the transmission of haptic information is the main issue in the context of the TI, as finding a suitable haptic codec is a challenging issue for transmission over the Internet. Besides, a haptic codec may take other kinds of data into consideration, such as video and audio traffic that may be multiplexed with haptic data for compression and transmission over the same physical network.

The TI cannot come to light without close collaboration among telecommunication engineers, computer scientists and mechanical engineers. The telecommunication engineer should provide a suitable haptic communication with high reliability, while artificial intelligence (AI) is needed to improve the human–machine interaction, which is the basis of the TI. Mechanical engineers are needed to build the robots that interact with haptic instructions and achieve the specified task. To this end, the TI is also called the Internet of Skills, since haptic data can implicitly transfer a skill through the movement of hands, such as teaching a child to play the piano from the other side of the world, diagnosing a patient using a distant doctor, or practicing a chirurgical operation through a robot.

1.4. Cyber-Physical Systems and TI

The TI is fulfilling the requirements of the Cyber-Physical Systems (CPS), which were designed based on the interaction between hardware and software, including all of the algorithms and intelligent decisions to be taken in the system, in addition to the connectivity among these elements. In brief, it refers to the interaction between the real world and the information technology that designates the TI as a perfect candidate for CPS.

The matching between the TI and CPS is shown in Figure 1.4. In the following, we explain the components of the TI, by detailing all of the elements that constitute the TI: physical world, smart computing, Internet of Things, storage and computation, communication and feedback.

1.4.1. *Physical world*

The interaction of the TI with the physical world occurs through different use cases that share a common characteristic, which is characterized by controlling an object in a remote environment through the network. The use cases can include applications in the industry, especially in: automation; healthcare, represented by telesurgery and teleoperation where robotics are at the core of control process; virtual and augmented reality, with all of its different applications; gaming and entertainment and any application that makes the notion of smart cities viable, such as road traffic management using the cooperative driving of autonomous vehicles, smart management of resources, etc.

1.4.2. *Internet of Things*

IoT can be considered as an object with embedded sensors that collects data through sensing and actuating, which is then sent through the Internet to a predefined destination, in order to analyze it and make decisions in real-time or non-real-time, depending on the type of application. There is a variety of devices, software and communication protocols used to support the IoT functionalities. These functionalities should be carried out in an autonomous way, without human intervention. The TI is sometimes considered as an extension of IoT in a very large scale. This is due to the architecture of the TI, which incorporates IoT in both domains, while using 5G as the network domain. As a part of CPS, the interaction with the physical world in the TI is done through the IoT, which feeds the domains with several kinds of data.

1.4.3. *Communication*

The communication between the master and slave domains in the TI will play a great role in assuring the continuity of service between the master and slave domains.

Indeed, the networking technology connecting both domains has a great impact on the QoS requirements for the different applications of the TI. This is justified by the fact that each application has its own requirements, not only in terms of QoS, but also in terms of the QoE experienced by the user. Actually, the use of network technology is not restricted to the network domain. The master and slave domains may include a network of sensors, such as Zigbee nodes, which have the capability of sensing and sending data to the actuator. In turn, the actuator may be connected to the access network via Wireless Local Network Area (WLAN) technology, such as WiFi. However, the legacy generations of mobile and wireless networks may not meet the new requirements of QoS for TI applications; this is why 5G is identified as the privileged technology for access and core networks, due to its high data rate and the salient feature of its core network, represented by separating the control plane from the user plane with service-based oriented architecture. However, for critical use cases like teleoperation, a fast feedback is expected to ensure a good functioning of the whole system. In this case, FO is used to obtain a fast feedback and send huge amounts of data. As a result, the TI can be seen as a network of networks with specified data types and functionalities that suit haptic communications.

1.4.4. *Storage and computation*

Having storage and computation as one of the components of the TI is essential due to the nature of TI applications, which are mostly real-time and need fast treatment. The edges are composed of IoT devices that are sensing and actuating data in real-time. The data should be stored and processed and then transmitted through the network domain. The main issue is that the IoT devices have limited life duration with limited energy, also, some actions need a rapid reaction. Storing and processing may not be performed in the master domain where the devices have limited capability in terms of memory and computation. Therefore, integrating cloud computing into the architecture of the TI to assist in data management is necessitated. However, cloud computing is not that suitable for real-time application, due to the introduction of further delays in the communication. Hence, the idea of service proximity is introduced through the deployment of fog computing and edge computing that provide all the services, while being close to the users.

1.4.5. *Feedback*

The feedback in the TI is very important in ensuring the appropriate continuity of the service. However, depending on the type of the haptic data exchanged between the master and slave domains, the sort of feedback can be decided. In general, haptic communications can be categorized into two major classes, depending on the type of human interaction with the distant environment. The first one can be represented by passive communication; this means that the interaction between the human and the remote environment will be limited to the exploration of the objects. Perception is the main activity that can be considered; however, the manipulation of objects and surfaces is not covered (Antonakoglou *et al.* 2018). The type of feedback in this category is an open-loop feedback; only tactile information is exchanged, since the instructions are one-way commands to be sent, as shown in Figure 1.3.

The second category is active haptic communication that includes both the perception and manipulation of distant objects, depending on whether the remote environment is real or virtual. In this case, the closed-loop feedback permits the exchange of kinesthetic information, which needs an interactive cooperation between the master and slave domains. The slave domain is completely dependent on the master domain, as it waits for commands in order to act accordingly; at the same time, it sends feedback to the master domain to initiate the subsequent operations, knowing that the feedback cannot only denote signals or commands, but also audio and video feedback, depending mostly on the application used (Pai *et al.* 2018).

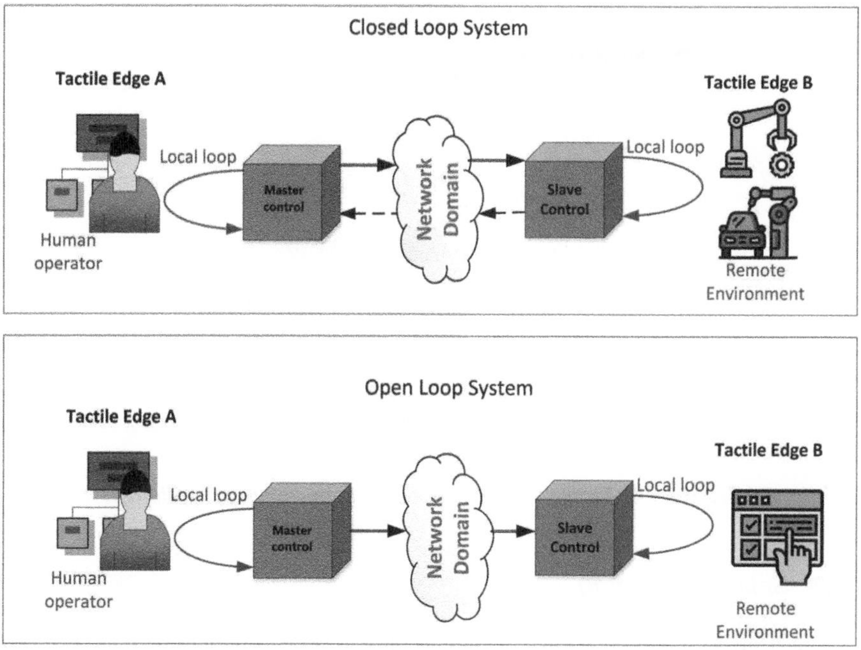

Figure 1.3. *Open-loop versus closed-loop systems. For a color version of this figure, see www.iste.co.uk/ali-yahiya/tactile.zip*

Figure 1.4. *Tactile Internet based on cyber-physical systems*

1.4.6. *Smart computing*

The need to make decision in order to ensure better predictions for the future, in an environment operating on the basis of action and reaction, is imperative. The difficulty is designated through leveraging data and making decisions in real-time. These decisions should be smart to ensure the awareness of the context of TI applications. Smart computing is dedicated for this purpose, with tools and methods that enable intelligent decisions to be made during ongoing sessions between master and slave domains. Exploiting the artificial intelligence (AI) concept, including all of its applications, can be used in the TI, especially when it concerns Machine Learning (ML) as its principal role in bringing intelligence to network domains through its prediction models. It is important to use ML and incorporate it in the TI architecture to solve the problem of stringent delay in haptic applications. Combining ML with the network part of the TI will play a significant role in giving the perception of zero-delay. This is done through the use of prediction models to predict movement on the remote domain of the TI. The prediction model will not only have an impact on the QoS, but also on the QoE of users who are involved in the loop. Investigating the appropriateness of a model for the TI context is a defying issue and should be selected well. Consequently, this would influence the precision of the decision of sending data from one domain to another.

1.5. References

Aijaz, A., Dawy, Z., Pappas, N., Simsek, M., Oteafy, S., Holland, O. (2018). Toward a tactile internet reference architecture: Vision and progress of the IEEE P1918.1 standard. *CoRR* [Online]. Available at: http://arxiv.org/abs/1807.11915.

Antonakoglou, K., Xu, X., Steinbach, E., Mahmoodi, T., Dohler, M. (2018). Toward haptic communications over the 5G tactile internet. *IEEE Communications Surveys Tutorials*, 20(4), 3034–3059.

Fettweis, G.P. (2014). The tactile internet: Applications and challenges. *IEEE Vehicular Technology Magazine*, 9(1), 64–70.

Pai, D.K., Rothwell, A., Wyder-Hodge, P., Wick, A., Fan, Y., Larionov, E., Harrison, D., Neog, D.R., Shing, C. (2018). The human touch: Measuring contact with real human soft tissues. *ACM Transactions on Graphics*, 37(4), 58:1–58:12.

ITU (2014). The Tactile Internet. ITU-T Technology Watch Report [Online]. Available at: https://www.itu.int/dms_pub/itu-t/oth/23/01/T23010000230001PDFE.pdf.

2

Reference Architecture of the Tactile Internet

Tara ALI-YAHIYA

Department of Computer Science, University of Paris-Saclay, France

The design of devices supporting haptic technologies is essential for delivering the new services and different use cases introduced by the TI. In addition, the network architecture should entail all the use cases and demonstrate the interaction among the different functional entities of the TI. Proposing a general architecture will accelerate the deployment of the TI as it prepares the basis for TI functioning and makes it easy to add or remove some elements that are not needed from one use case to another. The flexibility of the design of the architecture and its role to act as an umbrella, even for special use cases that can be branched from a broad one, is a challenging matter. From this concept, the working group of IEEE 1918.1 proceeded to define the specifications of the TI by describing a global architecture that includes all the entities, interfaces, codec protocols for different types of applications. In this chapter, we will describe the architecture proposed by IEEE 1918.1 with its related use cases, as a step towards understanding the big picture of the TI from a technical point of view.

2.1. Tactile Internet system architecture

The TI working group IEEE 1918.1 defined the baselines of an architecture that makes use of a variety of telecommunication technologies, proximity of the cloud computing services to the users and the intelligent predictive algorithms to make any

For a color version of all the figures in this chapter, see www.iste.co.uk/ali-yahiya/tactile.zip.

TI use case operational. It is worth mentioning that the TI is considered as an overlay paradigm that can be operated on different kinds of networks. Hence, the working group made all the efforts in order for its design to be flexible and compliant with the current ICT components. The IEEE 1918.1 architecture, as shown in Figure 2.1, is composed of three main parts, which are explicitly: (i) tactile edge A, which is normally called the master domain, (ii) network domain, which is responsible for traffic transport, and (iii) tactile edge B, which is called the slave domain. The master domain comprises tactile devices (TD) that may include different sensors and/or actuators (S/A) collecting data and forwarding or receiving them to/by the sensor/actuator gateway (SG/AG) that works as the interface between the S/A and the sensor/actuator nodes (SN/AN) (Aijaz *et al.* 2018; Holland *et al.* 2019). The TD can also include the human–system interface (HSI) node that produces haptic output, in this case it is called the HSI Node (HN) to convert the human inputs, such as instructions or voice commands into haptic inputs. Moreover, it can be represented by a controller node (CN) that provides control algorithms for the edge equipped with sensors and actuators such as sampling, multiplexing or demultiplexing for actuators and cognizance operations.

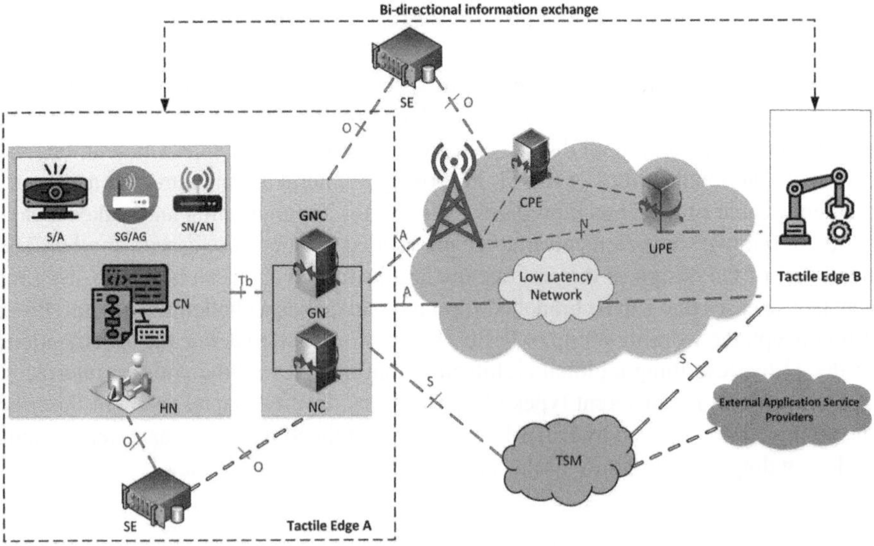

Figure 2.1. *TI reference architecture of IEEE 1918.1*

The main entity that interfaces the master to the network domain is the network controller (NC) and gateway node (GN) that can be collocated together in a Gateway Network Controller (GNC) either in the edge or in the network domain, as it is responsible for user plane data forwarding as well as control plane processing. The network domain is composed of a radio access network that is connected to the control plane entities (CPEs) and the user plane entities (UPEs) that reside in the core network.

An innovative node is added to the architecture, which is the support engine (SE) that ensures some functionalities such as caching, proximity computations and intelligent capabilities in order to ensure a reduced delay, especially for time-critical applications that require very low E2E latency. The SE can be located either in the master domain or in the cloud providing services to the network domain. In order to connect both tactile edges to the external application providers, the tactile service manager (TSM) will be responsible for the registration and authentication to that service in the third party network. One more interesting communication paradigm that is introduced in the IEEE 1918.1 is the method of connection between the two tactile edges via the network domain. Indeed, the latter can be composed of different types of networks, either wired, wireless or even a very fast and low latency network dedicated to low latency application traffic transportation such as fiber optic.

As for the interfaces, they are denoted by access (A), tactile (T_b), open (O), service (S) and network (N) interfaces, connecting the different elements of the architecture. For instance, the A interface connects the TD to the network domain through the GNC. While the T_b interface connects the entities inside one tactile edge. As for the O interface, it links the SE to other components of the architecture, whereas the S interface relays GNC to the TSM for sending control information and, finally, the N interface connects the internal entities of one network domain (Holland *et al.* 2019).

2.2. IEEE 1918.1 use cases

The IEEE 1918.1 working group has defined seven use cases that make use of the TI architecture in order to deliver its services. The tactile edge produces the haptic data as well as the other conventional Internet data. The production of application data depends on the type of use cases. These are identified by the IEEE 1918.1 standard, as shown in Figure 2.2:

1) teleoperation;
2) automotive;
3) immersive virtual reality (IVR);
4) Internet of drones;
5) interpersonal communication;
6) live haptic-enabled broadcast;
7) cooperative automated driving.

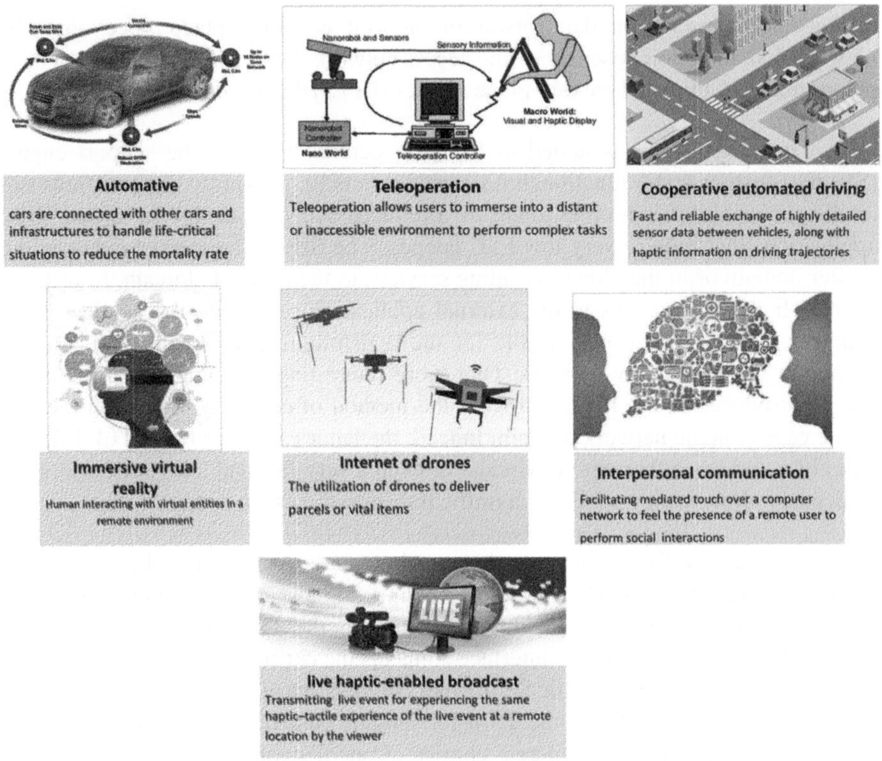
Automative
cars are connected with other cars and infrastructures to handle life-critical situations to reduce the mortality rate
Nanorobot and Sensors
Sensory Information
Macro World:
Visual and Haptic Display
Nano World
Teleoperation Controller
Teleoperation
Teleoperation allows users to immerse into a distant or inaccessible environment to perform complex tasks
Cooperative automated driving
Fast and reliable exchange of highly detailed sensor data between vehicles, along with haptic information on driving trajectories
Immersive virtual reality
Human interacting with virtual entities in a remote environment
Internet of drones
The utilization of drones to deliver parcels or vital items
Interpersonal communication
Facilitating mediated touch over a computer network to feel the presence of a remote user to perform social interactions
LIVE
live haptic-enabled broadcast
Transmitting live event for experiencing the same haptic-tactile experience of the live event at a remote location by the viewer

Figure 2.2. *Tactile Internet use cases*

2.2.1. *Teleoperation*

The teleoperation use case covers a wide range of operations that require controlling an object in an inaccessible remote environment to achieve some tasks. The best examples on teleoperation use cases are telesurgery and telerobotics (see Figure 2.3). Adding haptic modality to this system needs some components to be inserted, in order to provide a realistic perception about the environment from the point of view of users. The QoE of users should be as high as possible with regard to some predefined measurements. In order for this system to work properly, the operator or the user in the master domain would generate commands in terms of velocity and position to be transmitted through the network domain to the teleoperator in the slave domain. The teleoperator in turn would perform the instructions and send back feedback in the form of video, audio or haptic signals in order to close the global loop. The stability of the global loop is mainly bounded by the QoS as the delay requirement ranges between 1 and 10 ms so that the whole system works correctly.

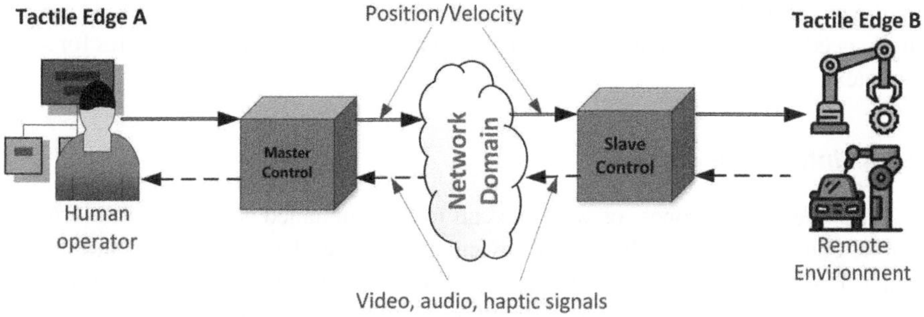

Figure 2.3. *Teleoperation use cases*

2.2.2. *Automotive*

Safety and security while driving a vehicle is not only related to the built-in functionalities of the car itself but also depends on the connectivity of the car with other cars. Cars of the future are expected to connect to each other and make appropriate decisions in good time. However, to make a good decision, the vehicle should be equipped with the necessary equipment that increases the interaction between the driver and the vehicle. From here, the two concepts that empower the TI, namely machine-to-machine and human-to-machine, will play a large role towards achieving this goal. For this purpose, sensors are used to collect data in real time combined with video and audio applications to help the driver in any critical situation. For the electronics of the car to communicate with each other, the IEEE 802.1 Time-Sensitive Networking (TSN) has proposed Audio Video Bridging (AVB) that connects the car's electronics through an Ethernet network adapted to the structure of the car. Haptic applications should be integrated to increase the interaction between the driver and the vehicle and the vehicle itself with other vehicles through the 5G network as the network domain, as feedback is important depending on the contextual information gathered by the vehicles.

2.2.3. *Immersive virtual reality (IVR)*

Virtual reality (VR) enables the user to interact with the computer environment through an interface that makes him/her feel like they are experiencing real life, as it engages all of their senses through simulation tools. Indeed, VR is today's killer application, including gaming and entertainment, training and simulations, transportation, manufacturing, and many other industry verticals. The term "immersive" specifies how users are involved in touching the virtual objects with all their senses. When the user interacts with distant virtual subjects through the network, the QoS in terms of data rate should be guaranteed. This is due to the greedy nature of VR traffic that requires a high data rate. The network plays a huge role in giving a real perception when dealing with haptic and tactile perception. Hence, a high-speed technology network such as FO and 5G could be good candidates for such a context.

2.2.4. *Internet of drones*

The Internet of drones, or what we call today unmanned aerial vehicles (UAVs), can be considered as the flying IoT devices that collect data from places difficult to reach. UAVs can have commercial, industrial and military missions, ranging from entertainment to surveillance to agriculture to parcel delivery. With the increasing number of UAVs, their traffic management should be taken into account as collisions may occur. From the TI perspective, the master domain may contain the dashboard that controls the drones in the slave domain through the network domain, as shown in Figure 2.4. However, due to the high speed of drones, again, the QoS in terms of low latency should be guaranteed. In addition, algorithms for drone route management should be implemented in an intelligent way to ensure the avoidance of collisions.

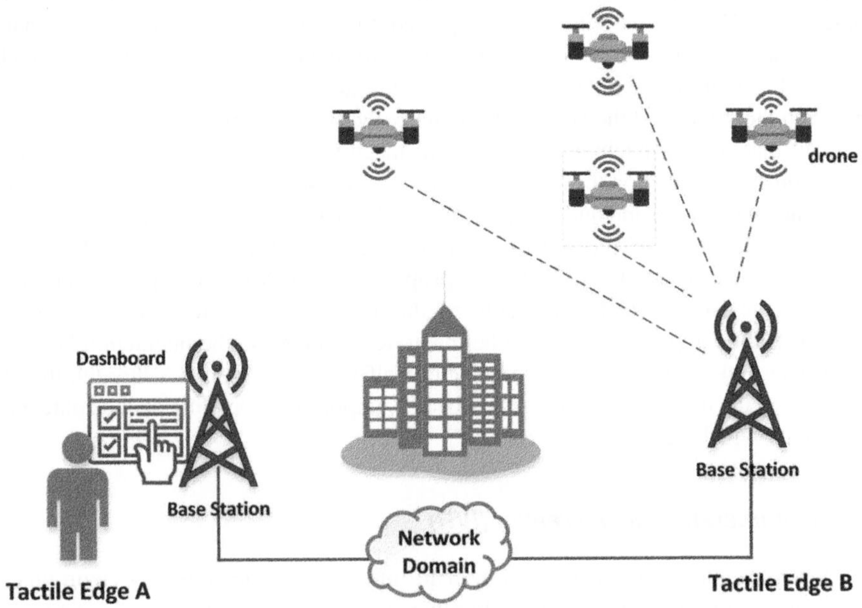

Figure 2.4. *Internet of drones use case*

2.2.5. *Interpersonal communication*

Interpersonal communication represented by audio call or any other form of Internet applications for delivering audio and video data is the basis for contact between people. However, such contact can lack feeling or emotion in remote communications. Moreover, for effective communication with touch and physical contact among family and friends, a haptic protocol needs to be used in the call. A haptic protocol for interpersonal communication requires that all haptic and non-haptic data be combined with each other. This means that both traditional Internet data and non-traditional, i.e. haptic, data should be sent at the same time in order to have a realistic feeling. It is important to mention that synchronization in haptic data exchange is very critical in such a context. Hence, QoE is prominent and user experience has a considerable effect on how the system works accurately (see Figure 2.5).

Figure 2.5. *Interpersonal communication use cases*

2.2.6. *Live haptic-enabled broadcast*

For a user to interact with a broadcasted event such as a football match or collaborative game played online, he or she can achieve it either passively or actively accordingly. Passively means that a user can only interact with the sense of vision and hearing, while actively means engaging other senses such as touching and feeling. The question is then how to deliver feeling or emotion to a distant user through TI technology from a broadcasted event. Incorporating haptics into the broadcasting of events requires frameworks that enable the integration of haptic effects into traditional audio-visual media. The difficulty in such platforms is to encode the haptic sensation and transmit it to the user who needs to be equipped with the appropriate electronic devices to decode signals that bring haptic effects to the users.

2.2.7. *Cooperative automated driving*

Self-driving vehicles or remotely controlled vehicles are the best case studies that can represent the TI relationship with vehicles. It can be represented by controlling a group of vehicles in an intelligent way for different reasons such as safety, security and traffic congestion. Cooperative automated driving has already been enabled by the vehicle-to-vehicle (V2V) or vehicle-to-everything (V2X) introduced in the latest release in LTE towards the path for 5G. The aim is to update all vehicles with regard to contextual awareness to have a complete vision of the roads and their capacities. This is done though protocols, algorithms for message dissemination among the vehicles and new haptic information exchange to support the TI paradigm.

2.3. Conclusion

Despite the general-purpose reference architecture proposed by IEEE1918.1, it is important to mention that this architecture can include any use cases and can even be personalized to special cases depending on the fine granularity of the application of the TI. The comprehension of the architecture would facilitate the understanding of the following chapters that rely mainly on the components of the architecture but from different points of view.

2.4. References

Aijaz, A., Dawy, Z., Pappas, N., Simsek, M., Oteafy, S., Holland, O. (2018). Toward a tactile internet reference architecture: Vision and progress of the IEEE P1918.1 standard. *CoRR* [Online]. Available at: http://arxiv.org/abs/1807.11915.

Holland, O., Steinbach, E., Prasad, R.V., Liu, Q., Dawy, Z., Aijaz, A., Pappas, N., Chandra, K., Rao, V.S., Oteafy, S., Eid, M., Luden, M., Bhardwaj, A., Liu, X., Sachs, J., Araúo, J. (2019). The IEEE 1918.1 "tactile internet" standards working group and its standards. *Proceedings of the IEEE*, 107(2), 256–279.

3

Tactile Internet Key Enablers

Tara ALI-YAHIYA

Department of Computer Science, University of Paris-Saclay, France

3.1. Introduction

The deployment of the TI cannot occur successfully without the support of some enabling technologies that ensure the ability to perform haptic operations. As illustrated in the previous chapter, the TI architecture is made up of innovative components that embody the network infrastructure, communication paradigm and cloud computing proximity to achieve high availability, low latency and security for the new use cases. In addition, some new softwarization techniques are based on micro-service concepts and virtualization for better management of physical resources in the network. As a step forward to realizing the TI, in this chapter we will discuss the TI-driven technologies that impact the haptic applications and services from the perspective of end-to-end communication architecture. These key technologies will assist in the convergence towards a common set of important design goals, allowing the TI to be realizable.

3.1.1. *The fifth-generation system architecture*

A key enabler of the TI is the 5G mobile network, considering that the legacy generations of mobile networks cannot meet the rigorous requirements of QoS of mission-critical haptic applications. For this reason, the 3GPP has brought new amendments to the specifications of 5G, known as New Radio (NR) 5G, to support new radio technology and to propose a new service-based architecture (SBA) that enables a flexible network configuration through the interaction of core network

The Tactile Internet,
coordinated by Tara ALI-YAHIYA and Wrya MONNET. © ISTE Ltd 2021.

functions with each other (Maier *et al.* 2016). A clear separation between the user plane and control plane is made to guarantee the scaling of resources in both planes independently. Both planes interact with each other via the use of a generic application programming interface (API) (Simsek *et al.* 2016).

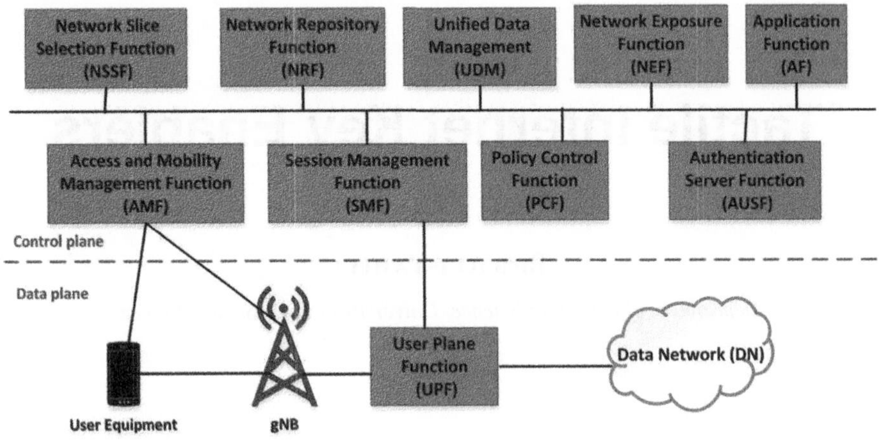

Figure 3.1. *5G architecture (ETSI 2018b). For a color version of this figure, see www.iste.co.uk/ali-yahiya/tactile.zip*

The architecture illustrated in Figure 3.1 describes the user equipment (UE) equipped with a 5G NR interface, connected to the next generation node B (gNB) as part of the Radio Access Network (RAN), while the key mobile core network functions can include an authentication server function (AUSF), an access and mobility management function (AMF), a network slice selection function (NSSF), a network repository function (NRF), a policy control function (PCF), a session management function (SMF), unified data management (UDM), a user plane function (UPF), an application function (AF) and a data network (DN) (ETSI 2018a).

The user plane includes the UPF, which is a fundamental component of the 5G architecture, as it is in charge of data forwarding and routing, connecting the components of the architecture to the DN, QoS handling and a mobility anchor among radio access technologies. The control plane, on the other hand, (referring to all the modules on the upper side of Figure 3.1) is composed of a platform that enables mobility management, access authentication and authorization, session management and exposes services to internal or external networks. This is done by means of a repository that contains all the services available in the network, in order that a network function registers either as a consumer or producer for a specified service. Each service in the network requires resources to be used; for that reason, the 5G NR core network provides a network slicing function that allocates a slice (a set of logical resources) to a service requested by a UE with predefined QoS requirements. All these functions, including network slicing, mobility, roaming, charging and control are managed through a unified policy rule platform, i.e. a PCF module.

The NR uses orthogonal frequency-division multiplexing (OFDM); however, the subcarrier spacing ranges from 15, 30 or 60 to 120 kHz which gives a variety of slot lengths for data transmission. This is very beneficial, especially when dealing with the diverse set of services in the TI that require different performances in terms of latency. Hence, when the duration of the slot decreases, the latency of transmission also decreases; choosing this configuration is very useful for the critical real-time application introduced in the TI. As for the QoS in 5G, it is flow-based, as the packets are classified according to the QFI (QoS flow identifier). 5G QoS flows are mapped in the access network to data radio bearers (DRBs), depending on the implicit requirements for each type of application, as shown in Figure 3.2.

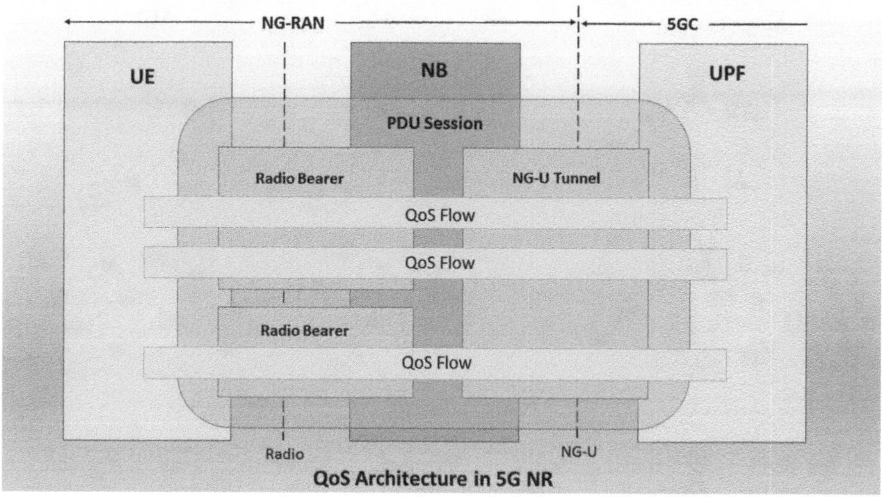

Figure 3.2. *5G QoS architecture. For a color version of this figure, see www.iste.co.uk/ali-yahiya/tactile.zip*

3.1.2. *Network slicing*

Network slicing is one of the foremost functionalities that is introduced in the 5G core network, ensuring that operators and telecommunication companies comply with the emerging requirements of enterprises. Enterprises nowadays provide different services that require different resources during a predefined time. Resources, here, can be represented by any network resources that are generally scarce, but, with the concept of network slicing, resources can be perceived as unlimited (Figure 3.3).

Network slicing can be defined as a set of logical networks being shared on the physical network infrastructure. Each logical network will be dedicated or assigned to a service of the TI that requires network resources for a duration of time. The network slicing is performing what is called the isolation of resources; this means that each slice will be independent and can have different types of resources (Ravindran *et al.* 2017; Rost *et al.* 2017; Zhang *et al.* 2017). The main aim is to meet the user's service level agreement (SLA) while increasing the revenue of the operator.

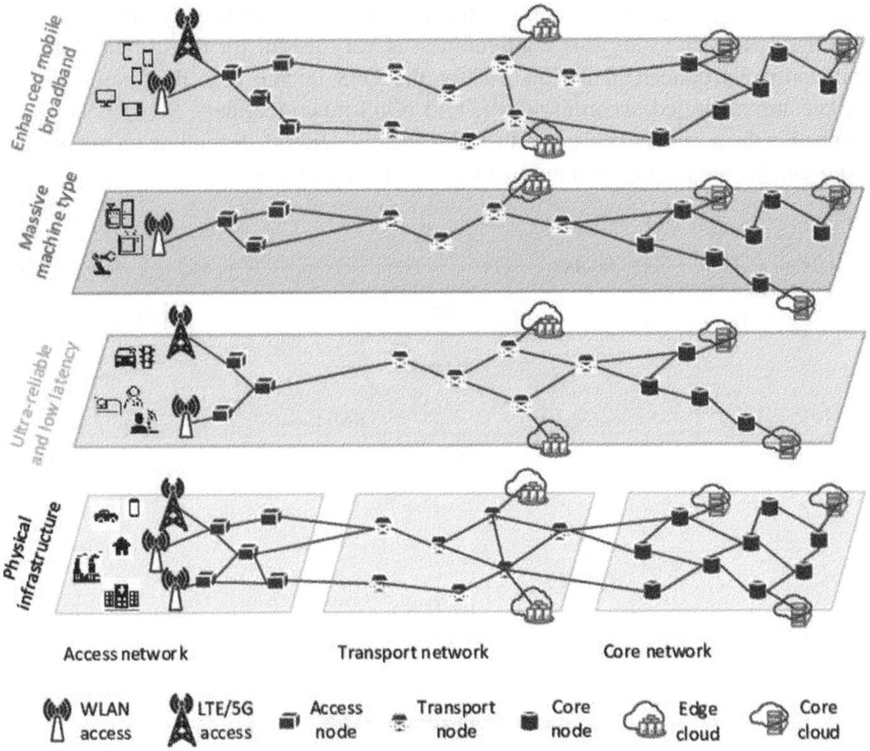

Figure 3.3. *Network slicing (Rost* et al. *2017). For a color version of this figure, see www.iste.co.uk/ali-yahiya/tactile.zip*

Network slicing is an effective method to optimize the resources in 5G by creating a virtual environment with the elements needed by any application. Its flexibility resides in its creation upon request and a lifecycle duration that lasts until the service ends and the slice is de-allocated. The logical network is instantiated through powerful techniques of virtualization, not only for the resources but also for the network functionalities related to these resources, and this is done by network function virtualization (NFV).

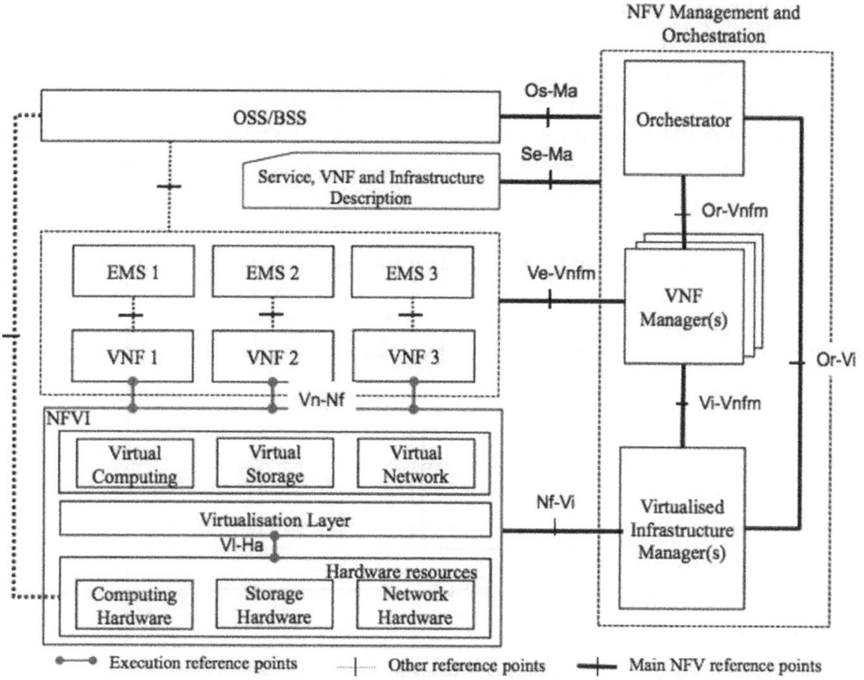

Figure 3.4. *Network function visualization (NFV) architecture*

3.1.3. *Network function virtualization*

While NFV tries to execute the functionalities of the network in a virtualized infrastructure, to provide the dynamicity and the adaptive configuration of the network, depending on the context that requires certain reconfiguration (ETSI 2013), the heterogeneous needs of the TI in terms of QoS requirements make NFV the best candidate to satisfy these requirements, especially in terms of latency, as its functionality can be adapted to the frequently changing nature of data and traffic, since tactile services may co-exist with non-tactile Internet services (Sachs *et al.* 2019). If this is the case, the traditional QoS mechanisms will fail to guarantee the QoS parameters and some mechanisms based on flow priority should be implemented to respect the end-to-end stringent delay of the TI traffic. Congestion solutions, load balancing, data path, physical location of the NFV, mobility tracking, etc.: all these functionalities will have an impact on the QoS guarantee and can be flexibly reconfigured to take into account the 1 ms delay of the TI (Giannoulakis *et al.* 2014). Finally, virtualization is an ideal technology to test new protocols and new architectures without stopping the operational network (Al Agha *et al.* 2016).

The advantages of the NFV for the virtualization technique are numerous. The first is being able to run on the same physical network, with the virtual networks using completely different technologies. One can, for example, get a first network using the IOS network operating system from Cisco, a second using JUNOS Juniper, a third and a fourth using the Nokia and Ericsson systems respectively, etc. It is also possible to run several different releases of the same network operating system. Obviously, the above examples have little chance of being realized because network manufacturers do not necessarily want to market their software without their equipment. Decoupling hardware and software will only be available for open systems (Al Agha *et al.* 2016).

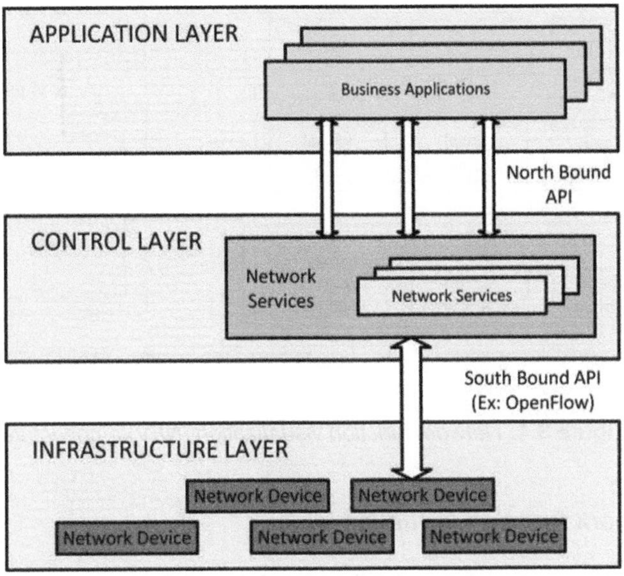

Figure 3.5. *Software-defined networking (SDN). For a color version of this figure, see www.iste.co.uk/ali-yahiya/tactile.zip*

3.1.4. *Software-defined networking*

Software-Defined Networking (SDN) is introduced to automate the network management. Indeed, all the functionalities of the hardware and network will be abstracted and softwarized to decouple the infrastructure from the functions. The SDN is aimed at separating the control plane from the delivery plane (data plane), so a centralized entity, i.e. the controller, can make the decision of delivery of the traffic according to actions corresponding to rules defined in the flow table of the controller (Costa-Requena *et al.* 2015; Ateya *et al.* 2018; Cabrera *et al.* 2019).

A protocol of communication should be defined between the control plane and the delivery plane, so the decision can be propagated from the controller to all devices supporting SDN. One of the famous examples is the OpenFlow, as it establishes a secure channel to monitor and configure the nodes located in the delivery plane. According to the context and the functionalities, i.e. routing, resource allocation, security configuration rules, etc., the controller would send its instructions to the network devices where the data plane resides, to be configured, as shown in Figure 3.6. SDN, NFV and network slicing are practical solutions that aid in the automation of the TI architecture while taking into consideration the specific need of each use case, defined by IEEE 1918.1.

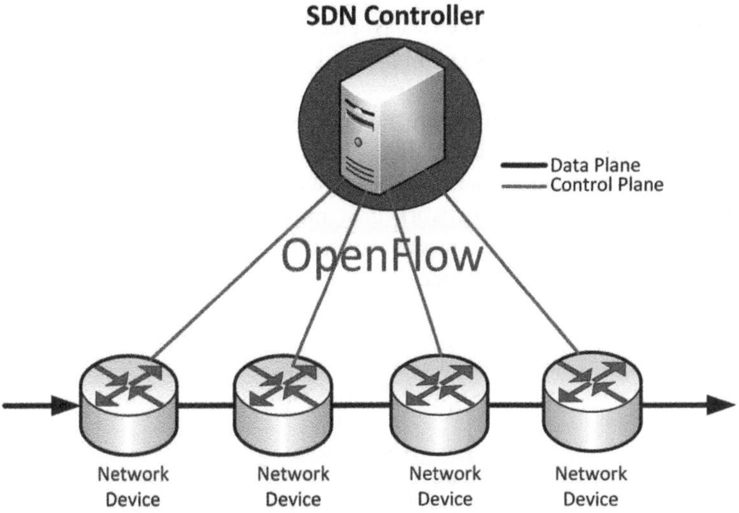

Figure 3.6. *SDN with openflow. For a color version of this figure, see www.iste.co.uk/ali-yahiya/tactile.zip*

3.1.5. *Edge computing*

Mobile edge computing (MEC) or multi-access edge computing tries to bring the cloud to the vicinity of the end-user, which enables the centralized cloud capabilities to be distributed to the edge of the radio access network. This paradigm assists in reducing delay, especially for the delay-sensitive applications while bringing the computation and processing closer to the users themselves (Tran *et al.* 2017). This would also assist in the case of the TI to enable users to track their real-time information, such as behavioral information, that can be predictive based on artificial intelligence algorithms.

The MEC is overcoming the classical cloud by having the edge device close to the user in order to compute, process and deliver instead of sending all the data through the backhaul. This will prevent congestion in which the latency, which the TI intends to guarantee, would collapse. The resource management can also be effectively achieved through a collaborative edge computation, as the task can be divided into small tasks and processed through different edge devices. This would be a good method of offloading the network (Mao *et al.* 2017). Providing a distributed caching mechanism by extending content delivery network (CDN) services towards the mobile edge can also enhance the users' QoE while reducing backhaul and core network usage (Skorin-Kapov *et al.* 2018). Indeed, there is a close relationship between the MEC and the QoE; however, since we are dealing with haptic applications, the codec used should reflect the combination between the tactile and kinesthetic. Consequently, a new design for mean opinion score (MOS) emerges, one that reflects users' best QoE through different codecs that also combines video and audio applications.

The TI specified, in its architecture, an entity called a support engine (SE), which stands for MEC functionalities for caching and processing. They specified its location in the master domain, where data can be collected from IoT devices nearby. This would mean isolating master network data movement from the core network or network domain and, as a consequence, guaranteeing a high delay of the real-time processing data needed by IoT devices.

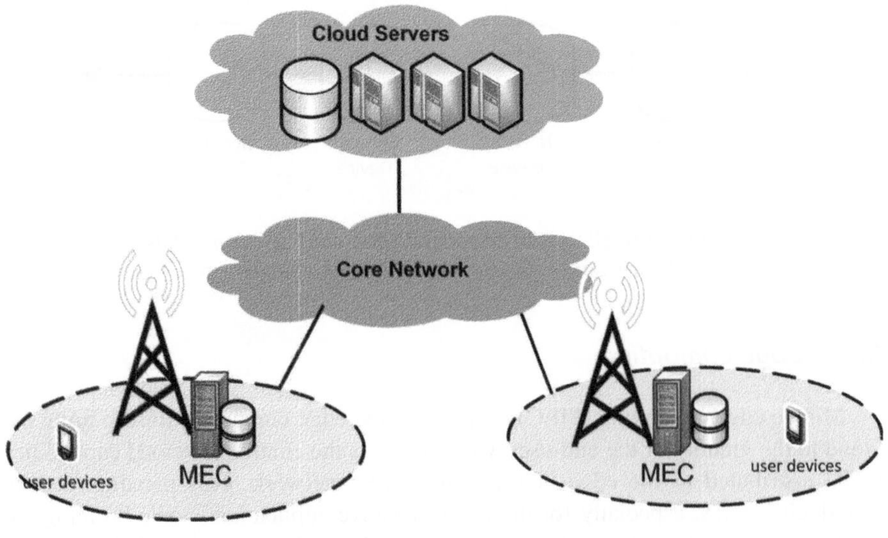

Figure 3.7. *MEC integrated with 5G. For a color version of this figure, see www.iste.co.uk/ali-yahiya/tactile.zip*

3.1.6. *Artificial intelligence*

The use of AI in applications requiring very low latency is highest when dealing with time-critical applications. AI with the type of network domain will determine the success of communication in the TI, especially for the teleoperation use cases demanding few milliseconds of E2E latency. The importance of latency lies in its effect on the stability of the whole system, as a small delay in delivering the data may cause incorrect manipulation at the teleoperator's edge. The latency is even more critical in active haptic communication.

The AI, from its definition, endows machines with intelligence, and plays an influential role in reducing low latency to give an environment with a perception of zero delay. The critical application that requires a 1 ms round trip delay requires a speed of light to transfer the packets; however, taking into consideration the distance, only 150 km can be traveled in both directions.

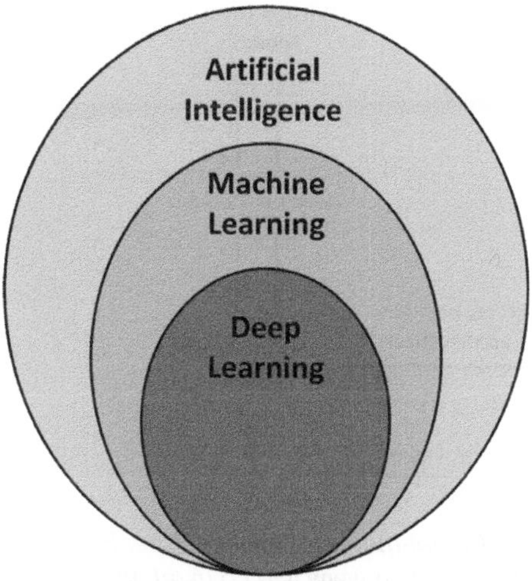

Figure 3.8. *Venn diagram of the relationship between artificial intelligence, machine learning and deep learning*

By deploying 5G, which is characterized by the high data rate and physical layer features that makes it possible for data to travel within a few milliseconds, it is still hard to deploy the mission-critical applications respecting the 1 ms latency (Morocho-Cayamcela *et al.* 2019). The network domain combined with AI tools can be the best way to overcome this problem. The AI, with its predictive models in ML, can be implemented in a different part of the network architecture, depending on the functionality that needs this tool. For example, the ML can be enabled in location-based services, caching and computation techniques in MEC, and its combination with SDN to control the traffic in the network. Additionally, networks nowadays tend to be autonomous, and lean towards self-organization without the intervention of humans.

AI as a whole is a vast domain; learning techniques are the most attractive applications to be used in the TI. Machine learning and deep learning fall under the broad category of AI (see Figure 3.8). To investigate their suitability for the TI, it is imperative to understand both of them. The ML techniques try to find a relationship between the input dataset and output actions to converge on an automation of the system without human intervention. An approach to proceed in ML is to break up the whole process into two phases: the training phase and the decision-making phase (Xie *et al.* 2019).

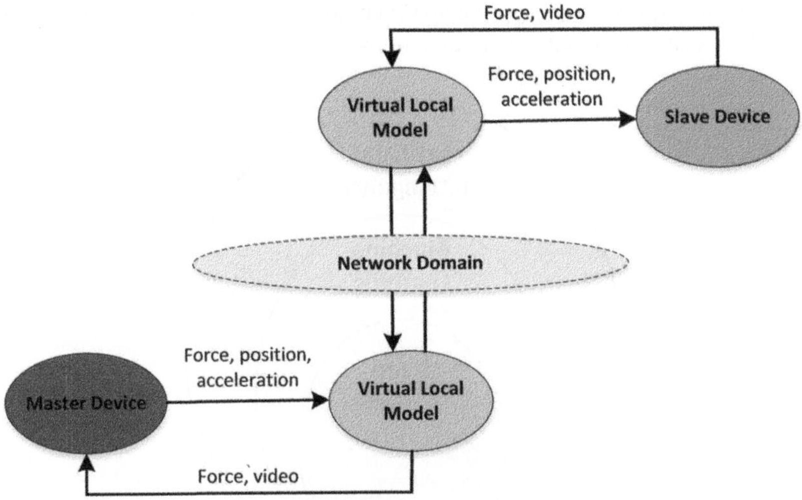

Figure 3.9. *Model-mediated architecture. For a color version of this figure, see www.iste.co.uk/ali-yahiya/tactile.zip*

According to the training dataset input to the system, the ML is applied, to learn the system model, while, in the decision phase, the output of the system is obtained based on the trained model. Machine learning algorithms are basically distinguished into four categories: supervised, unsupervised, semi-supervised and reinforcement learning.

Deep learning can be considered as a subset of ML; however, it tries to mimic biological nervous systems more and achieve representation learning, using multi-layer transformation. A deep learning prediction algorithm is composed of several layers. Normally, the input data passes through the layers in sequence, and each layer carries out matrix multiplications on the data. The output of a layer is usually the input to the subsequent layer. After data is processed by the final layer, the output is either a feature or a classification output.

The prediction models of ML can be integrated with the model-mediated system in order to predict the reaction of the remote teleoperation at the slave edge. The importance of the predictive models is driven from that fact that they have a critical impact regarding the stability of the whole system. This is why the model-mediated system is the best option when dealing with TI applications requiring very low latency. This model is based on providing a virtual version of the distant environment in the master domain, so the interaction will be in real-time, with different types of network domain. This interaction is based on a predictive model, based on ML, to predict the next movement of the teleoperator. In order to update the master domain and check its correctness and validity, the operator should continuously receive feedback from the teleoperation, in order to ensure that there is no force disturbance caused by any errors that occur (Ateya *et al.* 2019).

3.2. Conclusion

The key enablers described in this chapter are some technological advancements that can be considered as a step towards the deployment of a successful Tactile Internet use case. The combination of software and hardware involved in the deployment of the Tactile Internet's application may be the trigger to better understanding of how to build an application, and which key enablers to use. The usage of all key enablers is not necessary to have a successful use case; it depends all on the requirements and needs in terms of QoS, QoE and how each key enabler would contribute to the stability of the whole system.

3.3. References

Al Agha, K., Pujolle, G., Ali-Yahiya, T. (2016). *Mobile and Wireless Networks*. ISTE Ltd, London and John Wiley & Sons, New York.

Ateya, A.A., Muthanna, A., Gudkova, I., Abuarqoub, A., Vybornova, A., Koucheryavy, A. (2018). Development of intelligent core network for tactile internet and future smart systems. *Journal of Sensor and Actuator Networks*, 7(1), 1.

Ateya, A.A., Muthanna, A., Vybornova, A., Gudkova, I., Gaidamaka, Y., Abuarqoub, A., Algarni, A.D., Koucheryavy, A. (2019). Model mediation to overcome light limitations – Toward a secure tactile internet system. *Journal of Sensor and Actuator Networks*, 8(1), 6.

Cabrera, J.A., Schmoll, R., Nguyen, G.T., Pandi, S., Fitzek, F.H.P. (2019). Softwarization and network coding in the mobile edge cloud for the tactile internet. *Proceedings of the IEEE*, 107(2), 350–363.

Costa-Requena, J., Santos, J.L., Guasch, V.F., Ahokas, K., Premsankar, G., Luukkainen, S., Pérez, O.L., Itzazelaia, M.U., Ahmad, I., Liyanage, M., Ylianttila, M., de Oca, E.M. (2015). SDN and NFV integration in generalized mobile network architecture. *2015 European Conference on Networks and Communications (EuCNC)*, 154–158, Paris, France, June 29–July 2.

ETSI (2013). Network functions virtualization (NFV); architectural framework v1.1.1, Technical report, ETSI.

ETSI (2018a). 5G; NR; physical layer; general description (3GPP TS 38.201 version 15.0.0 release 15), Technical specification, ETSI.

ETSI (2018b). LTE; 5G; evolved universal terrestrial radio access (E-UTRA) and NR; service data adaptation protocol (SDAP) specification (3GPP TS 37.324 version 15.1.0 release 15), Technical specification, ETSI.

Giannoulakis, I., Kafetzakis, E., Xylouris, G., Gardikis, G., Kourtis, A. (2014). On the applications of efficient NFV management towards 5G networking. *1st International Conference on 5G for Ubiquitous Connectivity*, 1–5, Levi, Finland, 26–27 November.

Maier, M., Chowdhury, M., Rimal, B.P., Van, D.P. (2016). The tactile internet: Vision, recent progress, and open challenges. *IEEE Communications Magazine*, 54(5), 138–145.

Mao, Y., You, C., Zhang, J., Huang, K., Letaief, K.B. (2017). Mobile edge computing: Survey and research outlook. *CoRR*, abs/1701.01090 [Online]. Available at: http://arxiv.org/abs/1701.01090.

Morocho-Cayamcela, M.E., Lee, H., Lim, W. (2019). Machine learning for 5G/B5G mobile and wireless communications: Potential, limitations, and future directions. *IEEE Access*, 7, 137184–137206.

Ravindran, R., Chakraborti, A., Amin, S.O., Azgin, A., Wang, G. (2017). 5G-ICN: Delivering ICN services over 5G using network slicing. *IEEE Communications Magazine*, 55(5), 101–107.

Rost, P., Mannweiler, C., Michalopoulos, D.S., Sartori, C., Sciancalepore, V., Sastry, N., Holland, O., Tayade, S., Han, B., Bega, D., Aziz, D., Bakker, H. (2017). Network slicing to enable scalability and flexibility in 5G mobile networks. *IEEE Communications Magazine*, 55(5), 72–79.

Sachs, J., Andersson, L.A.A., Araúo, J., Curescu, C., Lundsö, J., Rune, G., Steinbach, E., Wikström, G. (2019). Adaptive 5G low-latency communication for tactile internet services. *Proceedings of the IEEE*, 107(2), 325–349.

Simsek, M., Aijaz, A., Dohler, M., Sachs, J., Fettweis, G. (2016). 5G-enabled tactile internet. *IEEE Journal on Selected Areas in Communications*, 34(3), 460–473.

Skorin-Kapov, L., Varela, M.V., HoBfeld, T., Chen, K.-T. (2018). A survey of emerging concepts and challenges for QoE management of multimedia services. *ACM Transactions on Multimedia Computing, Communications, and Applications*, 14(2s), 29:1–29:29 [Online]. Available at: http://doi.acm.org/10.1145/3176648.

Tran, T.X., Hajisami, A., Pandey, P., Pompili, D. (2017). Collaborative mobile edge computing in 5G networks: New paradigms, scenarios, and challenges. *IEEE Communications Magazine*, 55(4), 54–61.

Xie, J., Yu, F.R., Huang, T., Xie, R., Liu, J., Wang, C., Liu, Y. (2019). A survey of machine learning techniques applied to software defined networking (SDN): Research issues and challenges. *IEEE Communications Surveys Tutorials*, 21(1), 393–430.

Zhang, H., Liu, N., Chu, X., Long, K., Aghvami, A., Leung, V.C.M. (2017). Network slicing based 5G and future mobile networks: Mobility, resource management, and challenges. *IEEE Communications Magazine*, 55(8), 138–145.

4

6G for Tactile Internet

Pinar KIRCI[1] and Tara ALI-YAHIYA[2]

[1] *Department of Computer Engineering, Bursa Uludağ University, Turkey*

[2] *Department of Computer Science, University of Paris-Saclay, France*

Although 6G, or the 6th generation standard for wireless communication technologies, is in development, it is considered to be the key enabler for the TI. Not only will the technology provide high data rates to various applications that can be drivers for use cases of IEEE 1918.1 but it will also introduce intelligence into all the network layers. Since AI is the core functionality of some TI use cases that require the prediction of some reaction from the slave domain, 6G can be seen as the network domain of choice for the TI. In this chapter, we will explain the role of 6G as a new key enabler for the TI and its applications.

4.1. Introduction

The continuous improvement of network technology and the digitization of the economy are key drivers of 6G networks. Automated networks need paradigm changes in communication network management, where communication and computation should eventually merge. This will lead to an increase in user satisfaction, and will ultimately provide new, smaller domains for local and efficient service provisioning (e.g. private networks). Development efforts for 5G concepts, technologies and products converted the fabric of cellular networks with the introduction of eMBB, URLLC and mMTC for more dedicated services.

The novel technology areas in 6G will extend to enhanced mobile access and wireless backhauling with smart and reconfigurable metasurfaces. Here, the wireless channel will be designed to develop system performance and molecular

The Tactile Internet,

communications. A process that could be achieved using the TI for pairing network components, applying artificial intelligence (AI) for communications/networking and for managing network functionalities and operating improved security techniques for cyberattacks. Energy, health, mobility, transportation and manufacturing are examples of novel domains that future networks will be required to facilitate and support.

However, 5G is still undergoing expansion and may not have reached its full functionality, as outlined by standards, and now is the time to outline the milestones for basic 6G concepts. 6G may become the first network of its kind that will provide for humanity in emergency situations. With 6G, decisions for providing services will be intelligently presented with the use of a number of technology and data enablers: smart transportation, behavioral analysis stored in big data and health monitoring. The blockchain-based networks will redesign the architectures for network domains through the adoption of shared, immutable records of data transactions between varying parties.

Blockchains will develop the efficiency of trust in the communication area with a more robust system of authentication, helping to eliminate the risks of cyberattacks. In addition to this, the adoption of AI to administer the network lifecycle will lead to a substantial decrease in power consumption over network segments. This will reduce the influence of mobile networking over the environment and cut emissions, resulting in greener technologies. And it will show that communication engineering is not only for advancing standards of living but also for helping to preserve the world (David *et al.* 2020).

The future presents many technical challenges that the current 5G standard cannot meet; those problems will be solved by the next generation, i.e. 6G. High data rate is important but even more important is providing security (Al Mousa *et al.* 2020). To meet the needs of services and applications planned for post 5G and 6G networks, explorations are progressing towards the integration of architectures with the aim of supporting the variety of new computation-heavy and latency-sensitive applications in the context of the Tactile Internet. 5G deployments have restrictions in terms of integration of new applications. To deal with this problem, next-generation 6G systems worked on the convergence of technologies. 6G will present new system paradigms (e.g. human-in-the-loop communications and human-centric services) (Pérez *et al.* 2020). The mobile communication technique aims to provide ubiquitous connections between devices such as phones, laptops and buildings for the Internet of Things. 6G should provide for needs across many aspects such as latency mobility and connection density. For this reason, the high dimensions and abilities of 6G will help in the establishment of a connectivity ecosystem in the TI that focuses on building a remote and real-time interactive system; this will provide the ability to interact with real and virtual objects using wireless techniques (Gholipoor *et al.* 2020a; Jia *et al.* 2020).

The focus of the TI on ultra-low latency communications with high availability, reliability and security is regarded as a remarkable enabler of mission-critical IoT services. Tactile IoT presents a big change from conventional data-delivery networks to technology-transfer networks. However, current research on tactile IoT focuses on the point-to-point communication link that is able to connect a single haptic device. To reach the next level, multimodel data from massive distributed or area-based IoT devices with sporadic traffic are needed. This creates a huge challenge over the design of physical layer technologies because of the existence of multiple transmitters and limited radio resources. Thus, grant-free non-orthogonal multiple access (NOMA), which exploits the joint benefit of grantless access mechanism and non-orthogonal signal superposition, has been used to simultaneously realize low latency and high-efficiency massive access in tactile IoT in Ye *et al.* (2019).

The TI is one of the next-generation wireless network services with end-to-end (E2E) delay as low as 1 ms. This ultra-low E2E delay cannot be provided in the current 5G network architecture, and it is also vital to consider the delay of all parts of the network, including the radio access. However, heterogeneous services with variable requirements are proposed in the next-generation wireless networks, and these services are classified as enhanced mobile broadband (eMBB), massive machine-type communications (mMTC) and ultra-reliable and low latency communications (URLLC). eMBB services need high data rate, and mMTC services require a large number of connections with a low data rate for each connection.

The TI covers numerous applications, such as remote surgery, remote monitoring, distance education and remote driving. Because of the importance of E2E delay in TI services, it is crucial to pay attention to all delays in the network to provide the E2E delay requirement. Delays in an E2E connection are queuing delays, transmission delays and network processing delays. The required radio resources in the network are power and bandwidth. They are limited and it is necessary to distribute them to users according to their service requirements, such as data rate and delay. Also, the queuing delay and the transmission delay in the radio access are vital for delay-sensitive services, and they are influenced by the resource allocation scheme. Radio resource allocation (R-RA) has an important role in guaranteeing the QoS, and the TI is highly sensitive to delay; thus, R-RA becomes vital in this case (Gholipoor *et al.* 2020b).

4.2. The architecture of 6G

Since the 1980s, developments within information communication technology (ICT) have tended to result in next generation revolutionary technologies emerging in 10-year cycles, as shown in Figure 4.1. The continuous improvement of information and communication technology has played a crucial role in the perpetual development of the information system and the prosperity of society. Prior to 4G, mobile communication was concentrated on people-oriented individual consumer markets. 5G has achieved remarkable technological breakthroughs with better transmission speeds, ultra-low latency, reduced power consumption and a huge number of connections. After the integration of 5G and artificial intelligence, the progress made by next-generation information technologies – i.e. big data and edge computing – has promoted industry improvement in areas such as healthcare, manufacturing and transportation. The 6G network will assure that everything will be linked deeply, intelligently and seamlessly, to adapt to the deep integration of IoT in different industries (Lu and Zheng 2020).

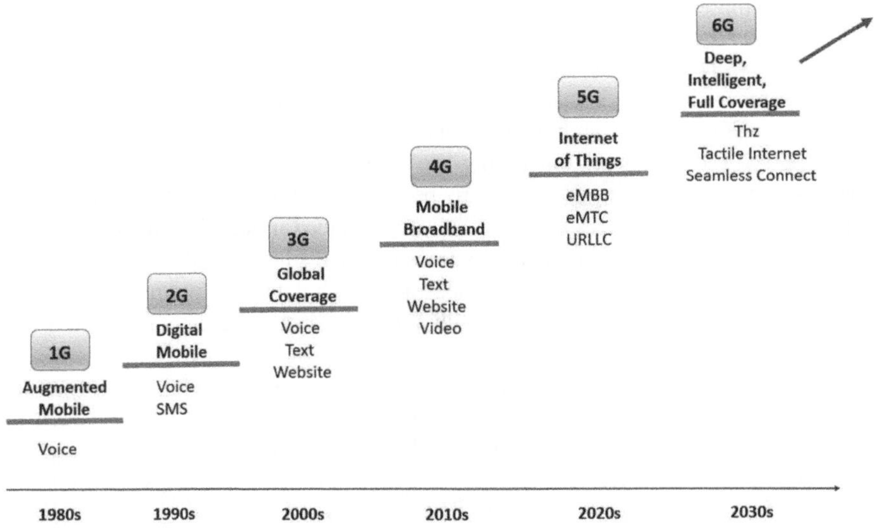

Figure 4.1. *Timeline of development trends in mobile communication*

4.2.1. *Network performance of 6G*

For the first time in ICT, improved performance will be demonstrated in the 6G system with visible light communication (VLC), wireless Tactile Internet (WTI) and high-performance computing (HPC). When compared with 5G, 6G will have high connection density, high peak rate, low latency, high user experience rate, high traffic density, strong mobility, strong positioning capability, high spectrum efficiency, strong spectrum support capability, high network energy efficiency, high reliability and so on. According to various perspectives of ICT, 6G will replace 5G and will be the next generation of communication systems. Table 4.1 shows the technological differences between 5G and 6G, focusing on the most important parameters in both generations.

	6G	5G
High peak rate	1 Tbps	20 Gbps
Experience rate	Counted as Gbps	Can reach 1 Gbps at most
Latency	As low as 0.1 ms	1 ms
Traffic density	100–10,000 Tbps/square meter	10 Tbps/square meter
Mobility	>1,000 km/h	500 km/h
Spectrum efficiency	200–300 bps/Hz	100 bps/Hz
Reliability (error coding rate)	Less than 1/1,000,000	Less than 1/100,000
Positioning capability	Outdoor/1 meter, indoor/10 cm	Outdoor/10 meter, indoor /1 meter
Spectrum support Capability	20 GHz/conventional carrier 100 GHz/multi-carrier aggregation	100 GHz/sub-conventional carrier 200 GHz/multi-carrier aggregation
Network efficiency	200 bits/J	100 bits/J

Table 4.1. *Comparison of 6G and 5G (Lu and Zheng 2020)*

A potential architecture of 6G is presented in Al Mousa *et al.* (2020) which has a space–air–ground–underwater integrated four-tier network. This 6G architecture features a satellite-based IoT rather than fiber optics and BS. Also, satellite launch and deployment are performed in space. Satellite communication has a significant and vital role in 6G, with the latter being known as 5G with a satellite network for some researchers. That is, the satellite network is associated with the basis of 5G to provide global coverage. Thus, 6G networks provide flexible and unlimited space communications to the users. The composition of 6G networks is shown in Figure 4.2.

4.2.2. *Space network*

A huge number of satellites are distributed in the space network. As for the different orbital altitudes of communication satellites, they can be divided into low Earth orbit (LEO) at 500–2,000 km, medium Earth orbit (MEO) at 2,000–36,000 km and high geostationary Earth orbit (GEO) at 36,000 km. In addition, communications between satellites in high orbits and those between high orbits and UAVs can use visible light lasers. The broadcast and multicast technologies can be used to improve the network capacity and decrease traffic burden in the multi-layered satellite communication system. Various services such as emergency rescue, Earth observation and navigation are ensured by the global coverage of the space network.

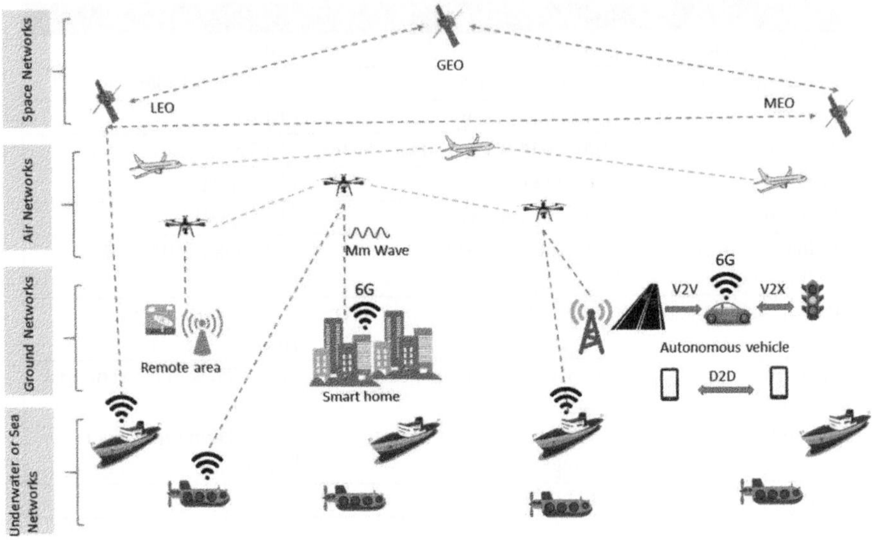

Figure 4.2. *Architecture of 6G. For a color version of this figure, see www.iste.co.uk/ali-yahiya/tactile.zip*

4.2.3. *Air network*

The air network is composed of different types of mobile aircraft such as UAVs, drones, balloons and airplanes. The mobile aircraft units under consideration are flying BSs which operate at varying altitudes. Aircraft at different altitudes or of different types may constitute aircraft-to-aircraft connections to provide services on the ground. In addition to this, the air network can be characterized as low cost with convenient deployment and wide coverage, which may provide regional wireless access services.

4.2.4. *Ground network*

Millimeter-wave (mmWave) and terahertz communications suffer from path loss, thus, the density of BSs will be high. 6G will use nano antennas, which are widely distributed, including in towns, at roadsides, airports and many other places. Thus, they allow people to use intelligent network services in remote areas. 6G can ensure seamless global coverage by combining with the aforementioned satellite communication network. Moreover, user services will be more intelligent. Also, mobile devices can provide device-to-device (D2D) communication directly with high information transmission rate.

4.2.5. *Underwater network*

An underwater network provides communication services for wide-sea and deep-sea activities that may improve the development of ocean observation systems. In addition to this, an underwater network uses all kinds of underwater communication. It involves submarines, unmanned surface vehicles (USVs), sensors and many more devices. Thus, an underwater network can construct an underwater collaborative operation network and realize intercommunication with the other three-tier networks (Li *et al.* 2020).

4.3. 6G channel measurements and characteristics

6G wireless channels exist at multiple frequency bands and in multiple scenarios, as illustrated in Figure 4.3. Moreover, the channel sounders and characteristics for each individual channel differ greatly. A comprehensive survey of variable types of 6G channels is given by grouping them under all spectra and global coverage scenarios.

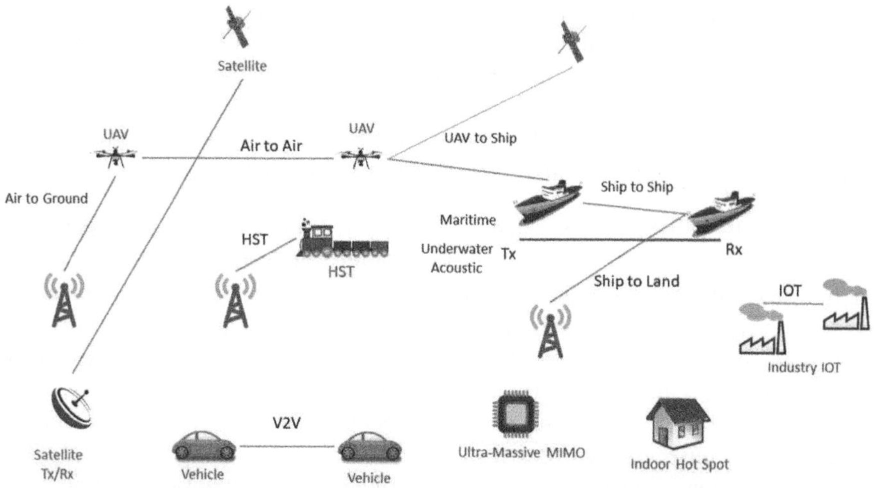

Figure 4.3. *6G wireless channels. Rx: receiver; Tx: transmitter. For a color version of this figure, see www.iste.co.uk/ali-yahiya/tactile.zip*

4.3.1. *Optical wireless channel*

Optical wireless bands refer to electromagnetic spectra with carrier frequencies of infrared, visible light and ultraviolet. They correspond to wavelengths in the range of 780–106 nm, 380–780 nm and 10–380 nm, respectively. They are used by wireless communications in indoor, outdoor, underground and underwater scenarios.

Optical wireless channels have unique channel characteristics. These are complex scattering properties for variable materials, nonlinear photoelectric characteristics at the transmitter/receiver ends, background noise effects, etc. The channel scenarios can be categorized as directed LOS, non-directed LOS and non-LOS (NLOS), tracked and so on. The basic difference among optical wireless and traditional frequency bands is that there is no multipath fading, Doppler effect or bandwidth regulation. Also, the measured channel parameters involve channel impulse response/channel transfer function, shadowing fading, path loss and root-mean-square (RMS) delay spread.

4.3.2. *Unmanned aerial vehicle (UAV) channel*

Utilization of UAVs has increased in recent years for both civil and military applications. The UAV channel also has unique characteristics: 3D deployment, spatial and temporal non-stationarity, high mobility and airframe shadowing, for instance. The UAV channel is categorized into air-to-air and air-to-ground channels. In general, two types of aerial vehicles are used for channel measurements, for example, small/medium-sized manned aircraft and UAVs. Channel measurements for the former are expensive, while the latter may lower the cost. Moreover, both the narrow-band and wide-band channel measurements have been conducted, most of which are at the 2-, 2.4- and 5.8-GHz bands. The evaluated environments involve urban, rural, suburban and open field. The evaluated channel parameters involve path loss, shadowing, fading, RMS delay spread, K-factor, amplitude and probability density function (PDF)/cumulative distribution function (CDF) (Wang *et al.* 2020).

Airborne base stations (BSs) that are carried by drones have great potential for augmenting the coverage and capacity of 6G cellular networks. Incidentally, one of the fundamental problems in the deployment of airborne BSs is the limited amount of available energy of a drone which shortens the flight time. The need to frequently visit the ground station (GS) to recharge limits the performance of the UAV-enabled cellular network and leaves the UAV's coverage area temporarily out of service. Drone-carried BSs are thought to be an integral part of the 6G cellular architecture. The inherent relocation flexibility and relative ease of distribution is useful for many requirements of next-generation cellular networks, i.e. ensuring coverage in hotspots and in areas with limited infrastructure, involving disaster recovery areas or rural areas.

The higher probability of providing a line of sight (LoS) link with ground users because of the high altitude leads to more reliable communication links and wider coverage areas. Possible use cases for airborne BSs include offloading macro BSs (MBSs) in urban and dense urban areas and providing coverage for rural areas that experience low cellular coverage due to a lack of incentives for operators. The air-to-ground (A2G) channel characteristics, optimal placement of UAVs and trajectory optimization are the most remarkable aspects of UAV-enabled cellular networks. Also, two key design challenges for UAV-enabled systems are discussed in Kishk *et al.* (2020). The first is the limited-energy resources available onboard that limits the flight time to less than one hour in most commercially available UAVs. The second is the wireless backhaul link.

4.3.3. *Underwater acoustic channel*

The underwater acoustic channel faces many problems. The applicable frequency is low and the transmission loss is high because of ambient noise in the oceans. Horizontal underwater channels are prone to multipath propagation due to refraction, reflection and scattering. The underwater acoustic channel scatters in both the time and frequency domains which causes time-varying and Doppler effects. Also, channel measurements were unusually conducted at several kilohertz, ranging from 2 to 32 kHz (Wang *et al.* 2020).

4.3.4. *Satellite channel*

Satellite communication has attracted significant attention in wireless communication systems. It is considered a novel solution to ensure global coverage due to its feasible services and lower cost. Satellite communication orbits are considered to be in geosynchronous orbit and non-geostationary orbit. The circular geosynchronous Earth orbit (GEO) is 35,786 km above Earth's equator and traces the direction of Earth's rotation. Non-geostationary orbits are classified as low Earth orbit (LEO), medium Earth orbit (MEO) and high Earth orbit (HEO). The classification depends on the distance of satellites from Earth. The applied frequency bands are the Ku (12–18 GHz), K (18–26.5 GHz), Ka (26.5–40 GHz) and V (40–75 GHz) bands. In general, the satellite communication channel is affected by weather dynamics, including rain, cloud, fog and snow. At frequency bands above 10 GHz, rain is the major source of attenuation. Moreover, the satellite communication channel presents large Doppler frequency shift, Doppler spread, frequency dependence, large coverage range, long communication distance and so on. The channel is viewed as LOS transmission, and multipath effects can be ignored because the distance is extremely long. High transmitted power and high antenna gains are required to fight against the high path loss caused by the long-distance and high-frequency bands (Wang *et al.* 2020).

Mobile communication standards have been improved for a new era of 5G and 6G. 6G will integrate with terrestrial mobile communications, high, medium and low Earth orbit satellite communications, and short-distance wireless communications. Moreover, 6G will combine communication, calculation, navigation, conception and intelligence. It will provide three-dimensional global coverage in space, on Earth and at sea with high-data-rate broadband communications with the use of intelligent mobility management.

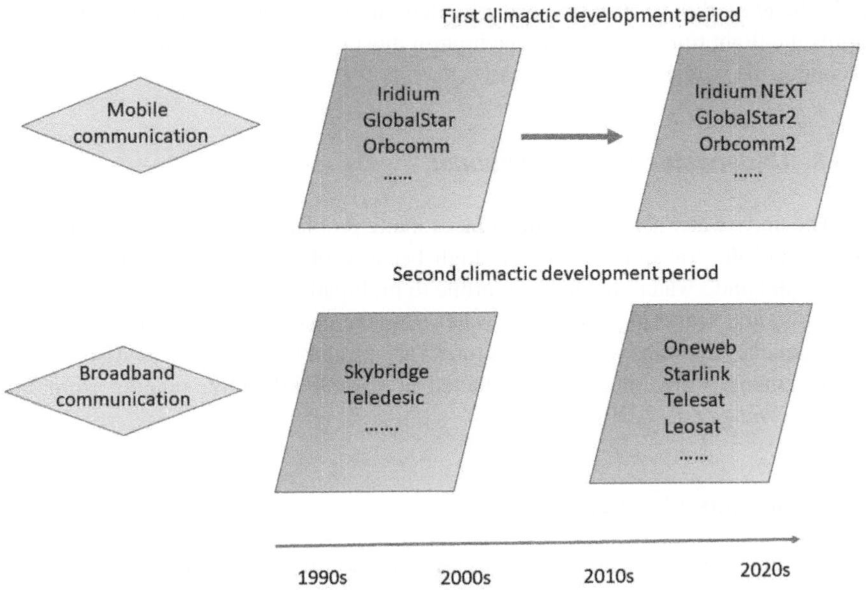

Figure 4.4. *Improvement periods for satellite communications*

The main aim of 6G is to provide seamless communication, anytime and anywhere. 6G will integrate with networks, terminals, frequencies and technologies, enabling wider innovation for ICT markets and applications. 6G will construct a universal mobile communication network. Research on 5G-based satellite communication will provide a basis for future 6G to integrate high, medium and low Earth orbit satellite communications and terrestrial communications.

In 6G, integrations cover the following six aspects:

– standard integration, a single standard supports both satellite mobile communication and terrestrial mobile communication;

– terminal integration, a UE has a unified identity for access. It is controlled uniformly by the network without distinguishing between the satellite network and the terrestrial network;

– network architecture integration, composed of a unified network architecture and a control management mechanism;

– platform integration, both satellite and terrestrial equipment use the same cloud platform architecture;

– frequency integration, terrestrial communication and satellite communication share spectrum by spectrum sensing, coordination, sharing and interference resistance;

– resource management integration, wireless resource management will be controlled and allocated uniformly, using terrestrial communication, satellite communication or both, for instance (Chen *et al.* 2020a).

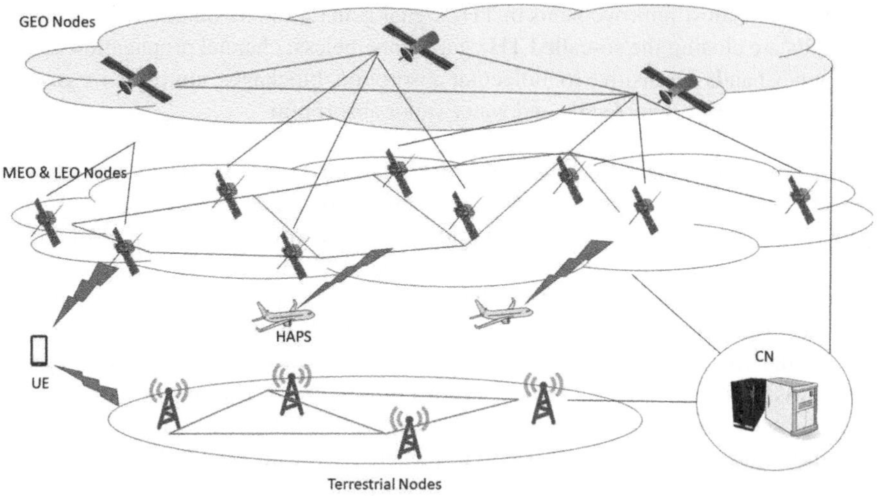

Figure 4.5. *System architecture of the 6G system. For a color version of this figure, see www.iste.co.uk/ali-yahiya/tactile.zip*

As illustrated in Figure 4.5, the 6G network will have a multi-layer architecture including user equipment (UE), satellite stations, terrestrial stations and core networks (CN). The satellite stations may include GEO, MEO and LEO. In addition, the terrestrial stations may have a macrocell base station, a microcell base station and a picocell base station. In the general architecture, both the satellite stations and terrestrial stations are seen as accessing nodes to communicate with UEs under the unified control and management of the CN. Also, satellite communication is compatible with 5G, developing 5G-based technologies and reusing 5G key technologies. However, satellite communication will be integrated within 6G, propagating at high, medium and low orbits and working with terrestrial mobile communication (Chen *et al.* 2020a).

4.3.5. *RF and terahertz networks in 6G*

High frequencies like terahertz (THz) will be central to 6G. Also, 6G networks will examine coexisting RF and mmWave deployments and coexisting RF and visible light communication (VLC) deployments. Moreover, higher frequencies in the terahertz (THz) band (0.1–10 THz) will be central to ubiquitous wireless communications in 6G. THz frequencies provide ample spectrum, above 100 Gigabit-per-second (Gbps) data rates, denser networks, massive connectivity and highly secure transmissions. The US National Science Foundation and the Semiconductor Research Consortium (SRC) identify THz as one of the four essential components of the next IT revolution. The THz spectrum is found between the mmWave and far-infrared (IR) bands. In addition, the latest improvements in THz signal generation, modulation and radiation methods are closing the so-called THz gap. Nonetheless, channel propagation at THz frequency bands is sensitive to molecular absorption, blockages, atmospheric gaseous losses due to oxygen molecule and water vapor absorption.

However, the conventional RF spectrum is characterized by strong transmission powers and wider coverage, but the spectrum is limited and congested. THz networks have reduced coverage so a trade-off exists between users' channel quality and available spectrum. To overcome the trade-offs between different frequencies, opportunistic spectrum selection mechanisms need to be provided with consideration for a coexisting network where RF BSs and THz BSs coexist. In a coexisting network, due to the enhanced signal power from RF BSs, the user will be biased towards RF SBSs. Despite the THz, BSs can ensure very large transmission bandwidth, yielding very high data rates and ultra-low latencies. In fact, new traffic offloading and user clustering schemes will be vital where users can be offloaded to different BSs and the resource utilization can be developed by balancing the traffic load among BSs (Hassan *et al.* 2020).

Recent evolutions in semiconductors have led to the terahertz band gaining notoriety as an enabler for terabit-per-second communications in 6G networks. Integrating this technology into complex mobile networks requires proper design of the full communication stack to address link and system-level problems related to network setup, management, coordination, energy efficiency and end-to-end connectivity. The 3rd Generation Partnership Project (3GPP) considers an extension to 71 GHz for 3GPP NR, as higher carrier frequencies come with larger bandwidth. Thus, the terahertz bands are considered to be a possible enabler of ultra-high data rates in sixth-generation (6G) networks. The spectrum from 100 GHz to 10 THz features vast chunks of untapped bandwidth for communication and sensing.

The IEEE has improved a physical layer that spans 50 GHz of bandwidth, between 275 and 325 GHz. Terahertz frequencies bring to the extreme the communications and networking challenges of the lower mmWave band. The harsh propagation environment features high path loss, inversely proportional to the square of the wavelength and to the size of a single antenna element, and, in addition, high molecular absorption in certain frequency bands. In addition, terahertz signals do not pass through common materials and are thus subject to blockages. Latest studies have focused on developing the communication range in macro scenarios and on signal generation and modulation. Directional antennas are used to decrease the increased path loss, because they can focus the power in narrow beams that increase the link budget, and to augment the security of wireless links, making eavesdropping more challenging. In addition, the small wavelength at terahertz provides many antenna elements to be packed in a small form factor (1,024 in 1 mm^2 at 1 THz). Therefore, it enables ultra-massive multiple-input multiple-output (UM-MIMO) techniques and array-of-subarrays solutions. Eventually, reconfigurable electronic surfaces can behave as smart reflectors and also overcome blockages in non-line of sight (NLoS).

A fraction of the hops between a client and a server will be on terahertz links. Integration into complex end-to-end networks should be considered, where many nodes and layers of the protocol stack interact to distribute packets between two applications at the two endpoints of a connection. Furthermore, the tough propagation characteristics of the terahertz band, the limited coverage of a terahertz access point, the directionality and the huge availability of bandwidth present new problems with a potential for the medium access control (MAC) network, transport layers, and also it may call for radical re-design of traditional paradigms for user and control planes of wireless networks. 6G will develop the energy efficiency of 5G networks, to equalize the higher number of terahertz nodes to be powered up than at mmWaves or sub-6 GHz.

Eventually, 6G networks will depend on a combination of sub-6 GHz mmWave and terahertz bands, and optical wireless links. The network infrastructure and mobile devices will need to adapt and use the carrier that ensures the best performance. 6G terahertz devices should take advantage of multi-connectivity, not only for the control plane and beam management but also for the user plane, forwarding data packets onto the various available radio interfaces to ensure diversity (Polese *et al.* 2020). The latest trend in alteration of wireless communications systems is towards higher data rates, system bandwidths, system capacities and operation frequencies. Furthermore, 5G is the first generation of mobile communication systems that provide millimeter-wave (mmWave) band transmission for high-speed wireless data transfer. It ensures transmission rates on the order of several gigabits per second using wide-transmission bandwidths up to a few hundred megahertz. Also, the next-generation 6G wireless communication systems are predicted to advance operations to upper mmWave band (100–300 GHz) and terahertz (THz) band (300–3,000 GHz) frequencies. Furthermore, future 6G systems will attempt to peak data rates up to terabits per second with low latency in transmission. Larger transmission bandwidths are required when compared to 5G systems and spectrum allocations below 100 GHz. For 6G systems, another fundamental physical restriction is the limited performance of electrical circuit technologies when converging the THz region.

Moreover, radio transceiver solutions for current 4G and 5G mobile terminals are implemented with complementary metal oxide semiconductor (CMOS) integrated circuit (IC) technology because of favorable cost, modularity and high level of integration. Nevertheless, mobile terminal power amplifiers (PAs) use either III–V technology, such as gallium arsenide (GaAs) or indium phosphide (InP), or silicon germanium (SiGe) heterojunction bipolar transistor (HBT) to overcome the restriction of CMOS and produce enough radio frequency power at frequencies below 6 GHz. In addition, the base station radios are using other III–V technologies in the power amplifiers and low-noise amplifiers (LNAs) to fulfill CMOS radio transceiver solutions in order to develop the radio performance. Besides, big and bulky discrete PA and LNA components are no longer applicable solutions with small antennas even in the lower mmWave region. Also, highly combined CMOS or SiGe HBT mmWave transceivers are required to be adopted as IC solutions with integrated PAs and LNAs next to the antennas minimizing form factor and any RF loss degrading the performance in phased arrays. Moreover, future 6G frequencies at 100 GHz and above will encounter a major problem because of the available transistor speed (such as fmax, i.e. maximum frequency to achieve power gain), especially in silicon-based technologies such as CMOS and SiGe HBT (Rikkinen *et al.* 2020).

4.3.6. *Visible light communication technology*

Visible light communication refers to a technology that uses light in the visible light band as an information carrier for data communication. Compared to radio communication, visible light has many advantages. Visible light can provide a lot of potentially available spectrum, and use of that spectrum does not need the authorization of spectrum regulators. Visible light does not produce electromagnetic radiation. It has many benefits, such as green environmental protection and no pollution. Thus, it can be widely used in places that house sensitive equipment, such as hospitals and gas stations, where electromagnetic interference can lead to serious problems. Visible light communication technology is high security because it cannot penetrate walls and other obstacles. Therefore, effectively preventing the transmission of information from being maliciously intercepted is possible, so the security of information can be provided. Today, many scientific research institutions in China, the United States, Germany, Italy and other countries have worked on visible light communication technology. The main bottleneck of this type of communication is caused by the state of visible light sending and receiving equipment. The modulation bandwidth of the transmitter presents waves greater than one millimeter only. However, the detector bandwidth and sensitivity are still very low. Furthermore, it is hard to provide the detection requirements in NLOS (non-line of sight) scenarios. Moreover, the terminal side requires control of the beam in order to realize a transceiver device by an integrated photonic antenna (Lu and Ning 2020).

4.3.7. *Orbital angular momentum technology*

Orbital angular momentum uses the orthogonal characteristics of vortex electromagnetic waves with variable eigenvalues to provide high-speed data transmission over the superposition of multiple vortex electromagnetic waves.

Thus, a new physical size for mobile communications is provided. Orbital angular momentum technology is categorized into two modes: the quantum state orbit and the statistical state. Nowadays, the application of orbital angular momentum in wireless communication involves many challenges. The industry has not gotten ahead of the miniaturization technology of orbital angular momentum, microwave quantum generation and coupling equipment; in addition, radio frequency statistical state orbital angular momentum transmission technology encounters the production of orthogonal vortex electromagnetic waves and vortices. Furthermore, detection and separation of eddy current electromagnetic waves, and how the influence of the transmission environment on these waves can be lessened, are two problems that are hard to solve (Lu and Ning 2020).

4.4. 6G cellular Internet of Things

According to International Data Corporation (IDC) predictions, by 2025 there will be 41.6 billion connected IoT devices, producing 79.4 zettabytes of data. Therefore, it is essential to form the sixth-generation (6G) cellular IoT network to meet the higher demands, such as wider coverage, increased capacities and ubiquitous connectivity. To provide real-time processing of mass data from terminal devices, 6G cellular IoT has to ensure accurate computation and efficient communication for a number of devices, which are considered to be two basic tasks of 6G cellular IoT. Besides, it is not unimportant to carry out the two tasks with limited wireless resources. For computation, the traditional way of transmitting and then computing is not appropriate for massive data aggregation in 6G cellular IoT because of the ultra-high latency and the low spectrum efficiency. To solve this problem, a novel solution called over-the-air computation (AirComp) is presented in Figure 4.3 that can compute the target functions including a summation structure over wireless multiple-access channels (MACs). AirComp decomposes the structure of the targeted function and then uses the superposition property of wireless MACs to achieve the sum result of pre-processing data by concurrent transmission. Eventually, the targeted function result can be achieved by post-processing the arrived signal at the base station (BS). In 6G cellular IoT, AirComp can integrate with multiple-input and multiple-output (MIMO) techniques to spatially multiplex multi-function computation and reduce computation errors by using spatial beamforming. For communication, conventional orthogonal multiple access (OMA) schemes cannot provide massive access due to the restricted radio spectrum. Non-orthogonal multiple access (NOMA) into cellular IoT is achieved to realize seamless access to a massive number of devices. Massive NOMA is exposed to co-channel interference that disrupts the quality of communication signals. Spatial beamforming is made use of to fight against co-channel interference and improve system performance. Since the BS of 6G cellular IoT will be equipped with a large-scale antenna array, there are ultra-high spatial degrees of freedom to reduce co-channel interference. To realize exact computation and efficient communication simultaneously, IoT devices should have sufficient energy. Besides, energy supply for massive IoT is an important task. Because of the high cost and environmental strain, frequent battery replacement for massive IoT is prohibitive. Thus, it is useful to adopt the wireless power transfer (WPT) technique to realize one-to-many charging by taking advantage of the open nature of the wireless broadcast channel (Qi *et al.* 2020).

4.5. Energy self-sustainability (ESS) in 6G

Following commercial deployment of 5G worldwide, studies of 6G mobile communication networks have increased. The major key performance indicator for 6G is its massive connectivity for small devices to enable the so-called Internet of Everything (IoE), a scale up from Internet of Things (IoT). Most of these IoE devices will be either battery-powered or battery-less, so the key challenge is how to prolong the lifetime of these IoE devices. Two aspects need to be addressed: energy efficiency to reduce the energy consumed by IoE devices, and energy self-supply to create new energy supplies for IoE devices. 6G will ensure energy self-sustainability (ESS) to massive IoE devices. 6G technologies such as THz and reflective/reconfigurable intelligent surfaces (RISs) have considerable potential to fulfill this vision of energy self-sustainability. THz frequencies are higher than mmWave and ensure better directionality in 5G, which is more efficient for wireless energy transfer (WET). RIS is applied in close proximity to end devices providing not only improved communication but also energy transfer to IoE devices, using either active or passive transmission. RIS may also be managed to provide on-demand WET (Yang *et al.* 2020b).

The main aim of 6G IoT networks is to ensure significantly higher data rates and extremely low latency. However, because of the scarce spectrum bands and ever-growing number of IoT devices (IoDs) deployed, 6G IoT networks face two vital problems, i.e. energy limitations and severe signal attenuation. Simultaneous wireless information and power transfer (SWIPT) and cooperative relaying are effective methods for solving both of these problems. 6G relies on massive collectivity where a large number of IoDs are connected for ubiquitous information exchange. However, most are powered by batteries with a restricted operating life. Energy harvesting (EH) is a convenient method for providing energy for 6G IoT networks. IoDs with EH technology can harvest energy from the environment, such as thermal, wind and solar, which means energy self-sustainability (ESS) can be achieved. Nevertheless, these energy sources may be unreliable because of the inherent unpredictability of the environment itself, which is a disadvantage of EH technology. Wireless power transfer (WPT) can ensure more reliable energy supply for IoDs compared to the EH obtained from radio frequency (RF) signals or from magnetic induction. Also, a new protocol is proposed for user cooperation in wireless powered networks to improve the energy efficiency. A carrier-sense multiple access with collision avoidance (CSMA/CA) protocol is proposed for WPT-enabled wireless networks to develop the throughput.

RF signals carry both energy and information; thus, simultaneous wireless information and power transfer (SWIPT) technology allows for energy harvesting and information decoding from the same received RF signal. Energy consumption of SWIPT IoT systems is reduced through two resource and task scheduling strategies. Because of the high-frequency bands used in the 6G IoT network, the transmitted signals are likely to be attenuated; thus, energy may not be harvested effectively at IoDs. Cooperative relaying can provide reliable communication and expand the coverage of IoT networks, a novel solution for solving the signal attenuation problem.

Furthermore, two relay destination selection SWIPT schemes for multi-relay cooperative IoT networks are proposed in Lu *et al.* (2020). Cooperative relaying-based SWIPT can effectively improve energy harvesting efficiency. Orthogonal frequency division multiplexing (OFDM) is mostly used because it can transmit signals efficiently over flexible subcarriers and power allocation. By combining OFDM and the SWIPT technology, a higher transmission rate and more efficient energy transfer in IoT networks can be achieved (Lu *et al.* 2020).

The main problem of the scalable deployment of the energy self-sustainability (ESS) Internet of Everything (IoE) for sixth-generation (6G) networks is balancing massive connectivity and high spectral efficiency (SE). Cell-free massive multiple-input multiple-output (CF mMIMO) is considered to be a novel solution where multiple wireless access points apply coherent signal processing to jointly serve the users. Besides, massive connectivity and high SE are hard to achieve at the same time because of the limited pilot resource. In 6G, the expectation is to obtain energy self-sustainability (ESS) Internet of Everything (IoE) networks, which handle massive connectivity and distribute large amounts of data traffic, while ensuring a more uniform quality of service (QoS) over the whole wireless network. Cellular massive multiple-input multiple-output (MIMO) is a known component of 5G networks. By inheriting several virtues from cellular massive MIMO, cell-free massive MIMO (CF mMIMO) is thought to potentially meet the needs of ESS IoE networks in 6G, strong macro-diversity and ubiquitous coverage, for instance. The main aim of CF mMIMO is for a large number of wireless access points (APs), which could be deployed in the coverage area and connected to a central processing unit (CPU), jointly serve all user equipment (UE) on the same time–frequency resource under the coordination of the CPU. Besides, an ESS network can be applied by wireless power transfer (WPT), where the ambient and dedicated radio frequency (RF) energy is harvested for the battery-limited UEs. Moreover, time-switching protocol is well known in WPT cellular massive MIMO operating in time division duplex (TDD) mode, where the transmission interval is partitioned into slots for energy harvest and information reception. The main idea behind integrating CF mMIMO with WPT is that each UE can be served by at least one AP with higher channel gain compared with cellular technology, which makes WPT applicable in the dense scenarios (Chen *et al.* 2020b).

4.6. IoT-integrated ultrasmart city life

The concept of a smart city is a city that improves the quality of life (QoL) of the people in it by optimizing its operations using available infrastructures to monitor, observe, examine and act by connecting the core components that run the city. With 5G, a city can be partially smart, which means major components such as healthcare, monitoring and transportation networks are individually smart. Also, there are many partially smart utilities such as electricity, water and waste. 6G will consider a holistic structure in an integrated way for a smart city. Some use cases about future city life are given below. Super smart home environment: Most of the elements will have a mastermind that will make decisions based on data fusion from a myriad of sensors embedded within them. The IoT infrastructure will be controlled by voice, gestures and other types of sensory communications. Thus, a complete overhaul in the approach to system structure will be required.

Ultra smart transport infrastructure: The automotive and transportation industries are attempting a change, partly because of the connectivity and networking capability presented by 5G systems and beyond. With 6G, the whole transport system will be affected by three factors. In-vehicle sensors and actuators will be intelligent, controlled by mastermind-like AI capabilities, leading to fully autonomous vehicles. Overcoming physical barriers will reduce most travel, and trips will be taken by driverless and fully automated vehicles. With the help of AI and extreme data rate capability, transport infrastructure will be fully autonomous where safety and security of the system will be provided by integrated intelligent sensors and actuators. A large amount of data will be required to be shared between vehicles to update live traffic and real-time hazard information on the roads and for high-definition 3D maps. Since vehicles move at very high speeds, the network requires a short round-trip time for communication. Vehicle-to-everything (V2X) technologies will not improve with 5G. Thus, its whole potential will be realized in 6G with VLC, OAM and emerging terahertz technologies.

Smart ubiquitous healthcare: The senescent population adds a massive cost to the healthcare system because of the continuation of conventional physical/manual administration and devices with limited communication and networking capabilities as well as restricted agility. The fast and anticipated developments in electronics and nano-bio sensors will improve health monitoring and management. The ubiquitous high-quality coverage of 6G will allow remote healthcare management intended to guarantee care regardless of the position of the patients. Latency and reliability should be guaranteed in order for emergency procedures and medical interventions to be given in a timely and uninterrupted manner. In addition, the improvements in soft and medical robotics, coupled with IXR, will mean remote surgery and medical intervention will be possible. Surgeons with particular expertise will be able to support and supervise robots to perform procedures from anywhere in the world as a result of vital latency and safety requirements being met. Privacy is a vital concern that should be primarily addressed by blockchain or variants of distributed ledger technology (Tariq *et al.* 2020).

Industry X.0: Efficient integration of robots with automation and warehouse transportation is crucial for industry growth. The concept of Industry X.0 is about improving Industry 4.0 by taking advantage of SMAC (social, mobile, analytics and cloud). Radio environments with a very complex network, composed of many robots, sensors and hardware elements is a problem. 6G will promote the Industry X.0 revolution by providing extreme latency and reliability as well as IoT and built-in AI capability (Tariq *et al.* 2020).

Pervasive AI: AI and its variants will be located at the core of 6G and will act as its vital enabling technology. Because of the improvements in AI techniques, deep learning and the availability of huge training data, there has been interest in using AI for the design and optimization of wireless networks. AI is expected to play an important role in a number of areas. These can be categorized into three levels, namely: in the device or end user equipment, localized network domain level and overall network level. This will convert the 6G network from a self-organizing regime to a self-sustaining one. At present, the most powerful AI technique is deep learning. It is based on a deep neural network (DNN) that relies on training in a centralized manner. 6G is proceeding towards a more distributed architecture like fog-RAN that provides billions of end-to-end communications anywhere around the world. The distributed cloud structure requires training to be conducted at the network edges. With 6G, AI should associate with game theory to provide a distributed learning mechanism where multiple AI agents can teach and learn from each other by interacting. Collective AI is a related concept that has been presented to deal with the situation where multiple AI agents want to perform the same goal based on local training with no direct communication between the agents. Furthermore, AI and hardware will be co-developed to provide better integration. Progress in formal methods will mean the devices are capable of reprogramming themselves, enabling them to reconfigure their functionalities as required for the purpose (Tariq *et al.* 2020): DNN and URLLC in 6G.

In sixth-generation networks, URLLC will lay the foundation for the emergence of mission-critical applications with stringent requirements in terms of end-to-end delay and reliability. Studies on URLLC are based on theoretical models and assumptions. The model-based solutions provide useful insights but cannot be implemented in practice. The objective of 6G networks is to enable ultra-reliable low-latency communications (URLLC), the foundation for allowing multiple mission-critical applications with stringent requirements on end-to-end (E2E) delay and reliability; for example, autonomous vehicles, factory automation and virtual/augmented reality (VR/AR). Also, the E2E delay is much longer than the transmission delays in the air interface. Meeting the E2E delay and reliability needs in such highly dynamic wireless networks proposes unprecedented problems in 6G networks. Model-based methods alone cannot solve the problems affecting 6G networks. To gain tractable results, some ideal assumptions and simplifications are unavoidable in model-based methods. As a result, the achieved solutions cannot satisfy the quality-of-service (QoS) requirements in real-world networks. With data-driven deep learning, it is feasible to learn a wide range of policies for wireless networks. To implement deep learning to URLLC, well-established models and theoretical formulas in communications and networking will be useful. Integrating model-based and data-driven methods is a novel approach in 6G networks. A multi-level architecture is an architecture that provides device intelligence, edge intelligence and cloud intelligence at user level, cell level and network level, respectively. Also, deep transfer learning and federated learning are accepted in this architecture.

To improve an architecture that provides deep learning for URLLC, the following features of 6G networks need to be considered. E2E QoS requirement: 5G systems are divided into multiple cascaded building blocks. Consequently, the E2E latency and reliability needs of URLLC are barely satisfied. In 6G networks, the E2E QoS requirement needs to be met by adjusting the whole network according to the stochastic service requests, queue states of buffers, workloads of servers and wireless channels. Scalable and flexible control plane: with a software-defined network, the control plane and user plane are slitted in 5G networks. To provide better scalability and flexibility in 6G networks, the network functions in the control plane can be fully centralized, partially centralized or fully distributed. So, the deep learning algorithms can be centralized or distributed depending on the network functions. Multi-level storage and computing resources: in 6G networks, storage and computing resources will be located at MUs, MEC and central cloud. The central cloud has many resources for offline training, but the communication delay among MUs and the cloud is long. With MEC, it is possible to train DNNs locally, so the response time of the network becomes shorter. By deploying computing resources at MUs, every device can decide with its local information in real time. Such a feature allows us to improve deep learning at various levels (Ye *et al.* 2019).

Based on these features, a wireless network is considered in Ye *et al.* (2019). It consists of smart MUs, MEC servers at APs and a central cloud, as given in Figure 4.6. To better show the multi-level architecture, mobility and traffic prediction for each MU, scheduler design at each AP and user association in a multi-AP network are examined.

Device intelligence at the user level: with device intelligence, MUs can make decisions based on local predicted information, such as the state of traffic and mobility. The prediction reliability is vital for decision-making; thus, the prediction error probability needs to be extremely low. Edge intelligence at the cell level: a scheduler at an AP maps the channel state information and queue state information to resource allocation between various MUs. With edge intelligence, DRL can be used to optimize the scheduler. Cloud intelligence at the network level: user association schemes rely on the large-scale channel gains from MUs to APs, as well as the packet arrival rate of every MU. With cloud intelligence, a centralized control plane uses a DNN. Thus, it approximates the optimal user association scheme, which maps the large-scale channel gains and the packet arrival rates of MUs to the user association scheme (Al Mousa *et al.* 2020).

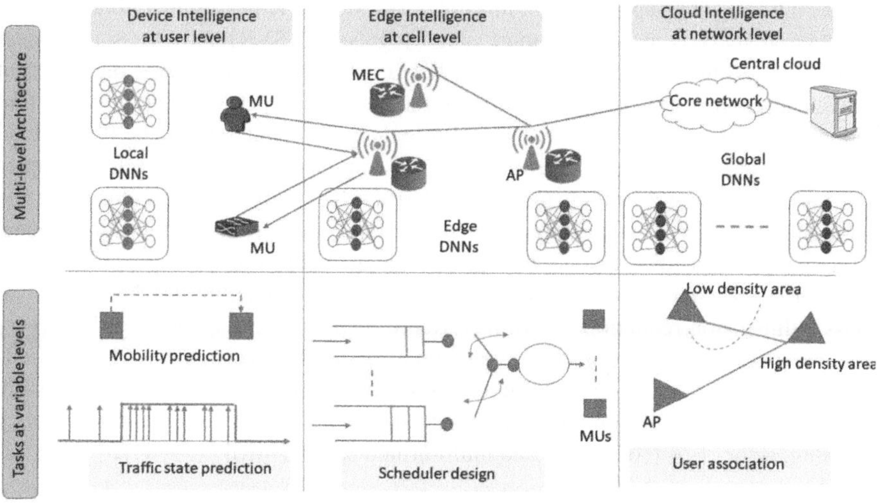

Figure 4.6. *Multi-level architecture in 6G. For a color version of this figure, see www.iste.co.uk/ali-yahiya/tactile.zip*

4.7. AI-enabled 6G networks

The development of 6G networks will be large-scale, multi-layered, highly complex, dynamic and heterogeneous. Furthermore, 6G networks should provide seamless connectivity and guarantee diverse QoS requirements of the huge number of devices. Also, they process a vast amount of data produced from physical environments. AI techniques with powerful analysis ability, learning ability, optimizing ability and intelligent recognition ability that can be used in 6G networks to intelligently carry out performance optimization, knowledge discovery, sophisticated learning, structure organization and complicated decision-making. With AI, an AI-enabled intelligent architecture is presented in Yang *et al.* (2020a) for 6G networks. It is made up of four layers: the intelligent sensing layer, the data mining and analytics layer, the intelligent control layer and the smart application layer, as shown in Figure 4.7. This four-layer bottom-up architecture can serve as a bridge between the physical world and social world. The physical world is composed of general physical/virtual things, objects, resources and so on. The social world is composed of human demand, social behavior, etc. Some common AI techniques are given below. AI techniques involve multidisciplinary techniques containing machine learning (supervised learning, unsupervised learning and reinforcement learning), deep learning, optimization theory, game theory and meta-heuristics. Supervised learning: supervised learning uses a set of exclusive labeled data to form the learning model, which is divided into classification and regression subfields. Classification examines aims to assign a categorical label to each input sample that contains decision trees (DT), support vector machines (SVM) and K-nearest neighbors (KNN). Regression analysis includes support vector regression (SVR) and Gaussian process regression (DPR) algorithms, and it predicts continuous values based on the input statistical features. Unsupervised learning: the aim of unsupervised learning is to find hidden patterns as well as extract the useful features from unlabeled data. It is divided into clustering and dimension reduction. Clustering seeks to group a set of samples into variable clusters according to their similarities. It contains K-means clustering and hierarchical clustering algorithms. Dimension reduction converts a high-dimensional data space into a low-dimensional space without losing much useful information. Also, principal component analysis (PCA) and isometric mapping (ISOMAP) are two classic dimension reduction algorithms. Reinforcement learning (RL): in RL, every agent learns to map situations to actions. It makes suitable decisions on which actions to take through interacting with the environment, to maximize a long-term reward. Classic RL algorithms contain Markov decision process (MDP), Q-learning, policy learning, actor critic (AC), DRL and multi-armed bandit (MRB).

Deep learning: deep learning is an AI function. It realizes the working of the human brain in understanding the data representations and creating patterns based on artificial neural networks. It is composed of multiple layers of neurons. Also, the learning model can be supervised, semi-supervised and unsupervised. Classic deep learning algorithms contain a deep neural network (DNN), convolutional neural network (CNN), recurrent neural network (RNN) and long short-term memory (LSTM).

Intelligent sensing layer: sensing and detection are the primitive tasks in 6G networks. 6G networks aim to intelligently sense and detect the data from physical environments through many devices such as cameras, sensors, vehicles, drones and smartphones or crowds of people. AI-enabled sensing and detecting can intelligently gather huge amounts of dynamic, diverse and scalable data by directly interfacing the physical environment, including radio frequency utilization identification, environment monitoring, spectrum sensing, intrusion detection, interference detection and so on.

Data mining and analytics layer: a core task that aims to process and analyze the huge amounts of raw data produced from the massive number of devices in 6G networks and provide semantic derivation and knowledge discovery. The data gathered from physical environments may be heterogeneous, nonlinear and high dimensional. Thus, data mining and analytics can be applied in 6G networks to solve the problems of processing the huge amount of data, as well as to examine the data collected towards knowledge discovery. Intelligent control layer: the intelligent control layer is composed of learning, optimization and decision-making. This layer uses the suitable knowledge from lower layers to provide multiple agents such as devices and BSs to smartly learn, optimize and choose the most suitable actions (e.g. power control, spectrum access, routing management and network association), with dual functions to encourage diverse services for social networks. This function is realized by applying AI techniques in 6G networks, where every agent is equipped with an intelligent brain (learning model) to automatically learn to make decisions by itself.

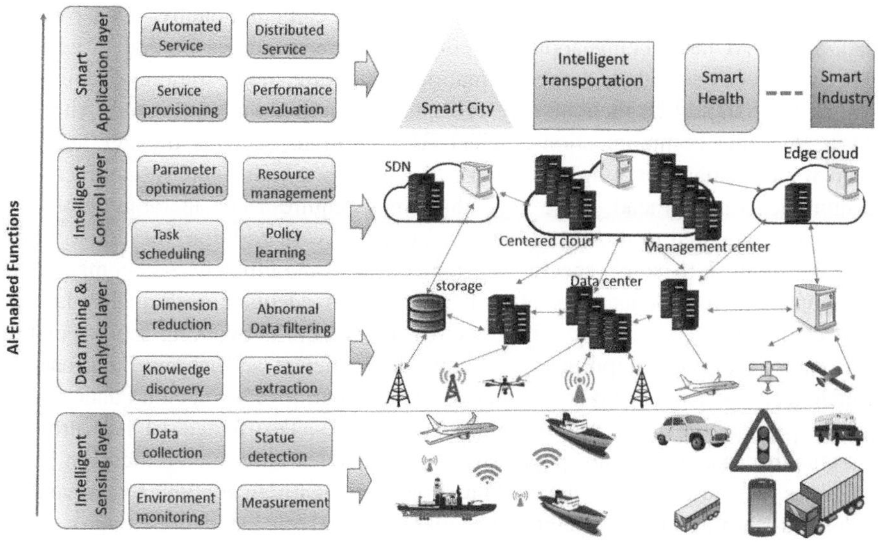

Figure 4.7. *AI-enabled intelligent 6G networks. For a color version of this figure, see www.iste.co.uk/ali-yahiya/tactile.zip*

Smart application layer: the aims of this layer are to deliver application-specific services to people according to their requirements and also to evaluate the provisioned services before feedbacking the evaluation results to the intelligence process. Intelligent programming and management can be provided by the impetus of AI to ensure more high-level smart applications, such as automated services, smart city, smart industry, smart transportation, smart grid and smart health, and handle global management relevant to all smart-type applications. All the activities of smart devices, terminals and infrastructures in 6G networks are administered by the smart application layer with the AI techniques to realize network self-organization ability (Yang *et al.* 2020a).

In particular, 6G will improve around ubiquitous AI (artificial intelligence), a hyper-flexible architecture that enables human-like intelligence for every aspect of networking systems. Edge intelligence is one of the important missing components in 5G. It is well known as a vital enabler for 6G to release the full potential of "edge-native AI" and bring network intelligentization. The influence of 6G in data collection, processing, transportation, learning and service delivery will design the progress of network intelligentization, catalyzing the maturity of edge intelligence. Next-generation edge-native AI technologies, in particular should provide the requirements below that are raised by 6G.

Resource-efficient AI: conventional wireless networks concentrate on augmenting the data transportation capability of wireless resources, spectrum and networking infrastructure, for instance. Besides, with more computationally intensive and data-driven AI tasks being applied by 6G, the extra resources needed to perform AI-based processes include data coordination, computing, model training, caching and so on. They should be meticulously evaluated, quantified and optimized. Although there are some communication-efficient AI algorithms such as transfer learning, deep reinforcement learning and federated learning, which demonstrate decreased communication overhead, these algorithms may require a considerable amount of resources compared to most data-centric applications. In addition, these algorithms may be applied to some specific learning tasks. Data-efficient AI: compared to computer vision systems, it is often hard to gather adequate high-quality labeled datasets under every possible wireless environment and networking setup. For this reason, it is of vital importance to shape data-efficient self-learning approaches, which need restricted or no hand-labeled data as input. Moreover, cloud data centers jointly provide the same set of computational tasks. Distributed AI has attracted attention because of the latest popularity of federated learning and its extension-based solutions. Besides, both distributed AI and federated learning continue to evolve. The federated-learning-enabled architecture will play a vital role in the future improvement of distributed AI-based 6G services and applications.

Personalized AI: personalized AI will play an important role in 6G to develop the decision-making abilities of AI algorithms, to assist machines to better understand human users' preferences and make better human-preferred decisions. Also, there are two types of human-in-the-loop AI approaches that include human intelligence as part of the decision-making process.

Human-oriented performance metrics: instead of concentrating on increasing traditional performance metrics, such as throughput, network capacity and convergence rate, the performance of 6G and AI should be jointly measured and considered by taking into consideration characteristics and potential responses of users. Furthermore, with 6G and mobile services becoming essential within human society, it is vital to improve new metrics, which can assist in providing the social and economic dimensions of 6G and AI convergence (Xiao *et al.* 2020).

4.8. AI- and ML-based security management in super IoT

5G cellular networks presented a new usage scenario to support massive IoT, i.e. mMTC. In terms of 6G, super IoT has been introduced, which can be enhanced with symbiotic radio and satellite-assisted IoT communications to support a massive number of connected IoT devices and provide extensive coverage. Eventually, more effective energy management mechanisms are expected to support the large scale of IoT systems which can then be used for long periods of time. Moreover, privacy and security issues will arise, especially for IoT systems gathering individual or sensitive information. As illustrated in Figure 4.8, AI and ML techniques are expected to assist 6G networks in making more optimized and adaptive data-driven decisions, in solving communication problems and in meeting the needs of emerging services.

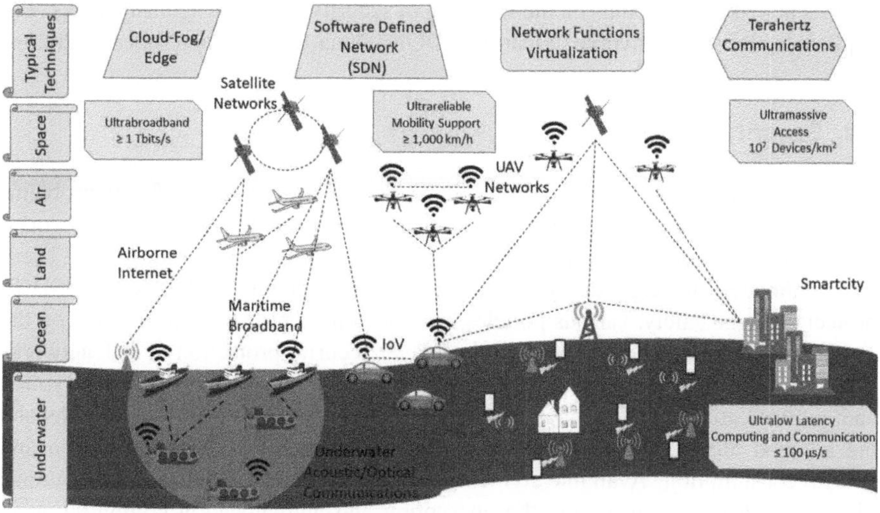

Figure 4.8. *AI/ML applications in 6G to support ultra-broadband, ultra-massive access and ultra-reliability/low latency. For a color version of this figure, see www.iste.co.uk/ali-yahiya/tactile.zip*

The vast amount of IoT devices and data cause significant problems in terms of privacy preservation and security guarantees. To preserve super IoT systems from various types of threats and attacks, authentication, access control and attack detection are of paramount importance. Conventional privacy and security technologies are barely feasible for super IoT because of the heterogeneity of resources, volume of networks, restricted energy, storage of devices and so on.

By supplying embedded intelligence in IoT devices and systems, AI/ML-based security technologies are used to deal with these security problems. There are many existing AI/ML-based solutions to address authentication, access control and attack detection in super IoT systems. Authentication and access control can assist IoT devices to notice identity-based attacks and hinder unauthorized devices from accessing authorized systems. To develop authentication accuracy, various AI/ML-based approaches can be well implemented based on various scenarios and assumptions. AI/ML technologies can be implemented to examine and notice various kinds of attacks including jamming, spoofing, denial of service (DoS) or distributed DoS (DDoS) attacks, eavesdropping and malware attacks, for instance. The supervised learning includes SVM, KNN, random forest (RF) and DNN, and it can be presented to distinguish these attacks by constructing classification and regression models. Furthermore, unsupervised learning can examine unlabeled data to partition them into several groups; for example, multivariate correlation analysis can assist in detecting DoS and DDoS attacks. Also, RL algorithms have been implemented to assist IoT devices in making decisions on the selection of security protocols against attacks. The applicable algorithms involve Q-learning, DQN, Dyna-Q and so on (Du *et al.* 2020).

4.9. Security for 6G

Data privacy and confidentiality will be one of the most significant problems for future 6G networks because of the rising network threats. In IoT networks, threats are composed of computer viruses, DDoS attacks and eavesdroppers. These threats threaten message safety and also copromise Quality of Service (QoS). To protect message safety, various policies have been presented that are focused on the physical layer and the link layer. The link-layer security protection of IoT networks is performed by the data authentication and encryption process. Available encryption and authentication methods have parameter configurations with the key lengths and the adopted algorithm, which result in varying levels of security protection and energy consumption. Available IoT chips support multiple security specifications. The main idea is to overcome the encryption and authentication configuration in the chip initialization process that facilitates the network configuration and fits the resource-constrained IoT chips. Besides, this security configuration may not ensure the service requirements of 6G networks for network QoS, energy efficiency and message safety. The network threats are mostly dynamic which may be far beyond the ensured low-level protection. The security protection causes extra energy consumption. A fixed high-level security configuration may soon run out of batteries, leading to the termination of service provision. It means that the fixed security configuration has the restriction of low energy efficiency. Taking into account the adopted energy harvesting techniques and service requirements in future 6G networks, the proper security configuration needs to be applied to the energy and network threats that can improve the security protection and network performance. Moreover, because of the network threats, the main strategy is to develop the security protection according to the energy. Also, when the harvesting source does not provide enough power, or the IoT devices have a huge sensing workload, it can consider the minimum level of required security protection to maximize working time. When the harvesting power is large, the security protection may be developed for better message safety. Another challenge is the sacrifice of network QoS caused by the extra communication overhead and energy consumption for security protection which is usually neglected. In the 6G IoT, network QoS is vital; thus, the sacrifice due to the security should be decreased (Mao *et al.* 2020).

4.10. The WEAF Mnecosystem (water, earth, air, fire micro/nanoecosystem) with 6G and Tactile Internet

Self-adapting capabilities, which 6G will push to the edge, urge for a pronounced separation between hardware (HW) and software (SW), this being in full contrast to the consolidated HW–SW co-design philosophy followed up to now. Algorithms will estimate the physical resources they can rely on in order to run in an optimal way. The HW will be required to achieve more symmetry with respect to the SW, in terms of flexibility, function reconfigurability and self-adaptivity/self-evolution. It envisions a prime role for micro/nanotechnologies, embodying materials, electronics and micro/nanosystems, in the scenario of 6G and of the Tactile Internet in Iannacci (2021). The WEAF Mnecosystem is designed using the analogy with the four classical elements in nature. Earth and air represent the classical concepts of HW and SW, respectively. Also, water is the new formulation of HW that, like water, is liquid in terms of functional characteristics. Furthermore, it achieves some features typical of the SW (i.e. air). Fire is the HW devoted to harvest, store and transfer energy, making it available everywhere at the network edge, i.e. when needed, where needed.

There are four identifiable application classes. They are labeled as verticals as they address top-down applications, i.e. from specifications to their physical (HW/SW) implementation. This plot is complemented by horizontals which are research activities, services and methodologies transversal with respect to verticals, yet functional to them and driven by their requirements. The visual representation of horizontals and verticals driven by MEMS within the 6G and TI scenarios is illustrated in Figure 4.9.

Verticals are grouped as follows, including but not restricted to the mentioned (MEMS) devices for 6G and TI applications:

– Telecommunications: highly reconfigurable and tunable broadband radio frequency (RF) passive components based on RF-MEMS/-NEMS technologies, like low-loss/high-isolation switches, switching matrices, as well as complex multi-state phase shifters, step attenuators, filters, resonators, impedance tuners, etc., for mmWaves (60 to 120 GHz) and THz (above 150 to 300 GHz);

– Energy conversion and storage: wideband microsystem-based energy harvesting (EH-MEMS) devices can transform environmental energy from mechanical vibrations, focusing at the IW (Indication Weights) range, examining various conversion mechanisms (piezoelectric, electrostatic, electromagnetic);

– Sensors and actuators: some indicative examples can be sensors (e.g. inertial, proximity, pressure) for gesture recognition and control of environments (domestic and industrial), sensors for BCI (Brain–Computer Interaction) and sensors for healthcare, entertainment and daily living support;

– Technologies for AI: with the rising trend for massive availability of small data, AI will be locally distributed down to the smaller pieces of HW (e.g. at sensor/component level), i.e. at the network edge.

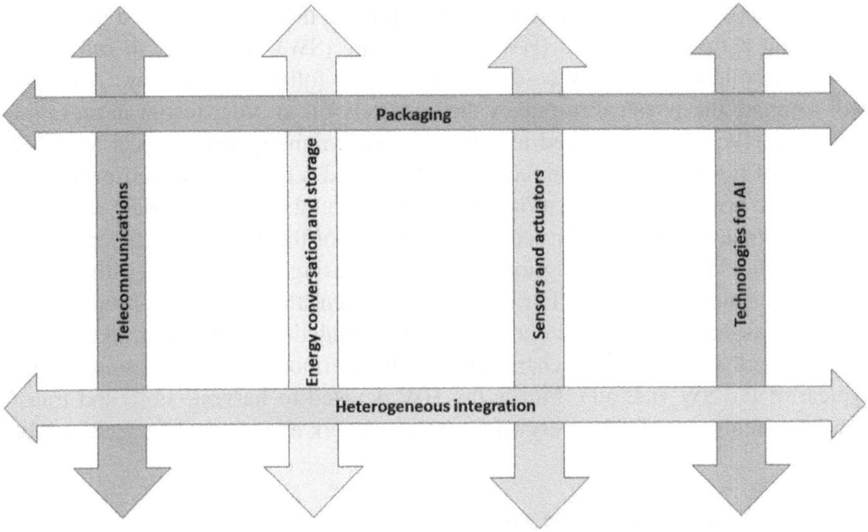

Figure 4.9. *Schema of vertical and horizontal MEMS-based application domains relevant to the 6G and TI future development and deployment. For a color version of this figure, see www.iste.co.uk/ali-yahiya/tactile.zip*

Horizontals are introduced as follows, and they fall across vertical application macro-domains:

– Packaging: development/extension of packaging and encapsulation technologies/methodologies for the physical devices spilling out from the above verticals decreasing the impact on their characteristics;

– Heterogeneous integration: development of integration solutions empowering the realization of hybrid HW sub-systems, for example, multi-source EH platforms (vibration, RF, thermal) interfaced to power management electronics and sensors/transceivers to be powered (Iannacci 2021).

4.11. References

Al Mousa, A., Al Qomri, M., Al Hajri, S., Zagrouba, R. (2020). Utilizing the eSIM for public key cryptography: A network security solution for 6G. *2020 2nd International Conference on Computer and Information Sciences (ICCIS)*, Jouf University, Sakaka, Saudi Arabia, April 7–9, pp. 1–6.

Chen, S., Sun, S., Kang, S. (2020a). System integration of terrestrial mobile communication and satellite communication – The trends, challenges and key technologies in B5G and 6G. *China Communications*, 17(12), 156–171.

Chen, S., Zhang, J., Jin, Y., Ai, B. (2020b). Wireless powered IOE for 6G: Massive access meets scalable cell-free massive MIMO. *China Communications*, 17(12), 92–109.

David, K., Al-Dulaimi, A., Haas, H., Hu, R.Q. (2020). Laying the milestones for 6G networks. *IEEE Vehicular Technology Magazine*, 15(4), 18–21.

Du, J., Jiang, C., Wang, J., Ren, Y., Debbah, M. (2020). Machine learning for 6G wireless networks: Carrying forward enhanced bandwidth, massive access, and ultrareliable/low-latency service. *IEEE Vehicular Technology Magazine*, 15(4), 122–134.

Gholipoor, N., Parsaeefard, S., Javan, M.R., Mokari, N., Saeedi, H., Pishro-Nik, H. (2020a). Resource management and admission control for tactile internet in next generation of radio access network. *IEEE Access*, 8, 136261–136277.

Gholipoor, N., Saeedi, H., Mokari, N., Jorswieck, E.A. (2020b). E2E QoS guarantee for the tactile internet via joint NFV and radio resource allocation. *IEEE Transactions on Network and Service Management*, 17(3), 1788–1804.

Hassan, N., Hossan, M.T., Tabassum, H. (2020). User association in coexisting RF and terahertz networks in 6G. *2020 IEEE Canadian Conference on Electrical and Computer Engineering (CCECE)*, London, Ontario, Canada, 30 August–2 September, pp. 1–5.

Iannacci, J. (2021). The WEAF Mnecosystem (water, earth, air, fire micro/nano ecosystem): A perspective of micro/nanotechnologies as pillars of future 6G and tactile internet (with focus on MEMS). *Microsystem Technologies* [Online]. Available at: https://doi.org/10.1007/s00542-020-05202-z.

Jia, M., Gao, Z., Guo, Q., Lin, Y., Gu, X. (2020). Sparse feature learning for correlation filter tracking toward 5G-enabled tactile internet. *IEEE Transactions on Industrial Informatics*, 16(3), 1904–1913.

Kishk, M., Bader, A., Alouini, M.S. (2020). Aerial base station deployment in 6G cellular networks using tethered drones: The mobility and endurance tradeoff. *IEEE Vehicular Technology Magazine*, 15(4), 103–111.

Li, W., Su, Z., Li, R., Zhang, K., Wang, Y. (2020). Blockchain-based data security for artificial intelligence applications in 6G networks. *IEEE Network*, 34(6), 31–37.

Lu, Y. and Ning, X. (2020). A vision of 6G – 5G's successor. *Journal of Management Analytics*, 7(3), 301–320.

Lu, Y. and Zheng, X. (2020). 6G: A survey on technologies, scenarios, challenges, and the related issues. *Journal of Industrial Information Integration*, 19, 100158.

Lu, W., Si, P., Liu, X., Li, B., Liu, Z., Zhao, N., Wu, Y. (2020). OFDM based bidirectional multi-relay SWIPT strategy for 6G IoT networks. *China Communications*, 17(12), 80–91.

Mao, B., Kawamoto, Y., Kato, N. (2020). AI-based joint optimization of QoS and security for 6G energy harvesting internet of things. *IEEE Internet of Things Journal*, 7(8), 7032–7042.

Pérez, G.O., Ebrahimzadeh, A., Maier, M., Hernández, J.A., López, D.L., Veiga, M.F. (2020). Decentralized coordination of converged tactile internet and MEC services in H-CRAN fiber wireless networks. *Journal of Lightwave Technology*, 38(18), 4935–4947.

Polese, M., Jornet, J.M., Melodia, T., Zorzi, M. (2020). Toward end-to-end, full-stack 6G terahertz networks. *IEEE Communications Magazine*, 58(11), 48–54.

Qi, Q., Chen, X., Zhong, C., Zhang, Z. (2020). Integration of energy, computation and communication in 6G cellular internet of things. *IEEE Communications Letters*, 24(6), 1333–1337.

Rikkinen, K., Kyosti, P., Leinonen, M.E., Berg, M., Parssinen, A. (2020). THz radio communication: Link budget analysis toward 6G. *IEEE Communications Magazine*, 58(11), 22–27.

Tariq, F., Khandaker, M.R.A., Wong, K.K., Imran, M.A., Bennis, M., Debbah, M. (2020). A speculative study on 6G. *IEEE Wireless Communications*, 27(4), 118–125.

Wang, C.X., Huang, J., Wang, H., Gao, X., You, X., Hao, Y. (2020). 6G wireless channel measurements and models: Trends and challenges. *IEEE Vehicular Technology Magazine*, 15(4), 22–32.

Xiao, Y., Shi, G., Li, Y., Saad, W., Poor, H.V. (2020). Toward self-learning edge intelligence in 6G. *IEEE Communications Magazine*, 58(12), 34–40.

Yang, H., Alphones, A., Xiong, Z., Niyato, D., Zhao, J., Wu, K. (2020a). Artificial-intelligence-enabled intelligent 6G networks. *IEEE Network*, 34(6), 272–280.

Yang, K., Jin, S., Rajatheva, N., Hu, J., Zhang, J. (2020b). Energy self-sustainability in 6G. *China Communications*, 17(12), iii–v.

Ye, N., Li, X., Yu, H., Wang, A., Liu, W., Hou, X. (2019). Deep learning aided grant-free NOMA toward reliable low-latency access in tactile internet of things. *IEEE Transactions on Industrial Informatics*, 15(5), 2995–3005.

5

IoT, IoE and Tactile Internet

Wrya MONNET

Department of Computer Science and Engineering,
University of Kurdistan Hewlêr, Erbil, Iraq

Connecting the physical world with the virtual one is done through sensors, actuators and computing units in the vicinity of the physical environment. This computing unit is usually small and is embedded in the physical device to be connected to the virtual one. The embedded system connected to the things provides them with intelligence. The resulting system is an Internet of Things (IoT). This has a wide range of applications in monitoring and controlling remote devices, automatically or through simple human issued commands. In automatic control, the machine, with the help of the sensors and actuators on the remote side, will manage the system. In the case of monitoring and interaction, the human is a part of the system, through the use of senses, such as sight, hearing, sent or touch. In addition, there may also be analytical tools provided by the Internet to process data and assist human interaction and decision. In this case, the resulting system becomes an Internet of Everything (IoE). A more particular case of the IoE is when the human is controlling a remote side, using the Internet as a communication medium, in order to become the master of the remote slave end. Then, the IoE system becomes a tactile one, more specifically the Tactile Internet (TI). In this case, the returned signal is mainly a tactile one, in addition to video or audio signals, which are used to control the remote side by the master. In this chapter, we review the types of IoT, classify them and show the existing architectures. Then, we will review the interaction models between IoT, IoE and TI.

5.1. From M2M to IoT

Machine-to-Machine (M2M) communication is considered as the foundation of the IoT. It started with wired connectivity between machines, using signaling to exchange information at a local level. The advent of computer networking and automation has allowed the use of M2M applications, such as supervisory control and data acquisition (SCADA). In contrast, the IoT uses Internet protocols to communicate data to the cloud or receive control commands from the IoT platform. The M2M requires reliable real-time communication between the machines, since critical applications such as machine automation are involved. In comparison, most IoT applications are less time critical as they are used for monitoring processes and device configuration or maintenance.

Nowadays, M2M converges towards IoT in terms of the communication network, by connecting distant automated teller machines (ATMs), security cameras and wind-farms. We can then admit that M2M becomes an integral part of the global IoT systems. In turn, the IoT has many common characteristics with the Tactile Internet (TI), such as Internet component connectivity and communication protocols. In the following sections, we will make the comparison by analyzing the IoT functioning, components and architectures.

	Monitoring-based	Control-based
Mission critical	– Low latency, – Carrier-grade reliability, – ~ 100% availability	– Ultra-low latency, – Carrier-grade reliability, – ~100% availability
Non mission critical	– Moderate latency, – Moderate reliability, – High availability	– Ultra-low latency, – Moderate reliability, – High availability

Table 5.1. *The matrix of IoT classification and the performance requirements (Zhang and Fitzek 2015)*

5.2. Classification of remote monitoring and control systems

For a better distinction between IoT and TI, it is useful to classify the IoT applications according to their performance and requirements. In Zhang and Fitzek (2015), the authors carried out this classification based on the orientation of the application: monitoring or control, and the criticality of the mission. Table 5.1 shows this classification with their performance requirements in terms of latency, reliability and availability. IoT's control-based applications are realized through simple messaging, such as switching an interrupter or a motor ON or OFF. A more complex system interfaces the digital world to the physical world through sensors and actuators that solve complex control problems. It falls under a much broader category called Cyber-Physical Systems (CPS) (Engineering 2011; Laszlo 2014). These systems contribute to security, efficiency and comfort in contrast to IoT systems' monitoring and simple actuation tasks. As such, the TI can be considered as a special case of CPS.

In the following sections, the constituents, architecture and protocols of IoT are presented. This presentation aims to analyzing the IoT for a broader comparison with the TI.

5.3. IoT-enabling technologies

Many technologies enabled the addition of the things (objects of the physical world) dimension to the Internet and its ubiquitous dimension. They can be regrouped into:

5.3.1. *IoT hardware*

The availability of processing power with small size and low power consumption, in addition to low prices, paved the way towards pushing processing power to the edge of the network. The increased processing power and memory size made it possible to embed the network stack on a small processor embedded in the physical objects, with sensors and actuators, chosen according to specific applications. Hardware technologies, such as application-specific integrated circuits (ASIC), field-programmable gate arrays (FPGA), smart sensors and the wireless physical layers protocols, such as IEEE 802.15.4 and 802.11, embedded on small processors, are the main hardware technology enablers for the IoT.

5.3.2. *IoT software*

Software development environments with open-source projects, with high-level language compilers for different hardware platforms, and the myriads of device libraries, helped accelerate the software developments for IoT systems. The open-source (hardware and software) platform Arduino is a real example of how the software environment with its simple utilization procedure, made IoT available for hobbyists, researchers and industrial communities. Moreover, the openness increases the community contribution, with an improved quality of code. Consequently, the faster development of IoT systems becomes possible.

5.3.3. *IoT connectivity*

Initially, radio frequency identification (RFID) was used to connect objects to the virtual world. Using both its active and passive tags, a connected reader scans the tags' data and sends it via the Internet to its destination. Later, with the appearance of Zigbee, WiFi and 6LOWPAN protocols, and their implementation on tiny processors, physical objects could be wirelessly connected to the Internet via a gateway. The low power consumption of the Zigbee protocol guarantees the autonomous functioning of the connected devices for years. This low power consumption is essential since the IoT should be able to cope with the number of connected objects' scale-up. New specialized wireless networks have emerged to connect things; these are wide-range wireless networks like NB-IoT, Sigfox, LoRa and LTE-M.

5.4. Architectural design and interfaces

Different authors have proposed many architectural designs (GrØnbæk 2008; Gardašević *et al.* 2017; Jamali *et al.* 2020). They are based on layers, ranging from three layers (application, network and perception) to five layers (business, application, processing, transport and perception) (Jamali *et al.* 2020). Figure 5.1 shows these two architectures.

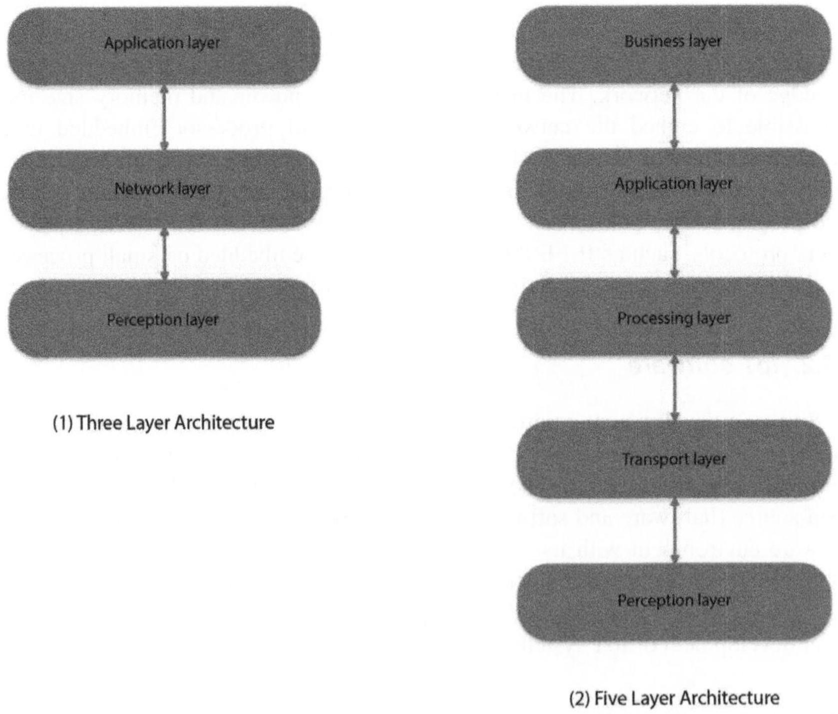

Figure 5.1. *Three- and five-layer IoT architectures. For a color version of this figure, see www.iste.co.uk/ali-yahiya/tactile.zip*

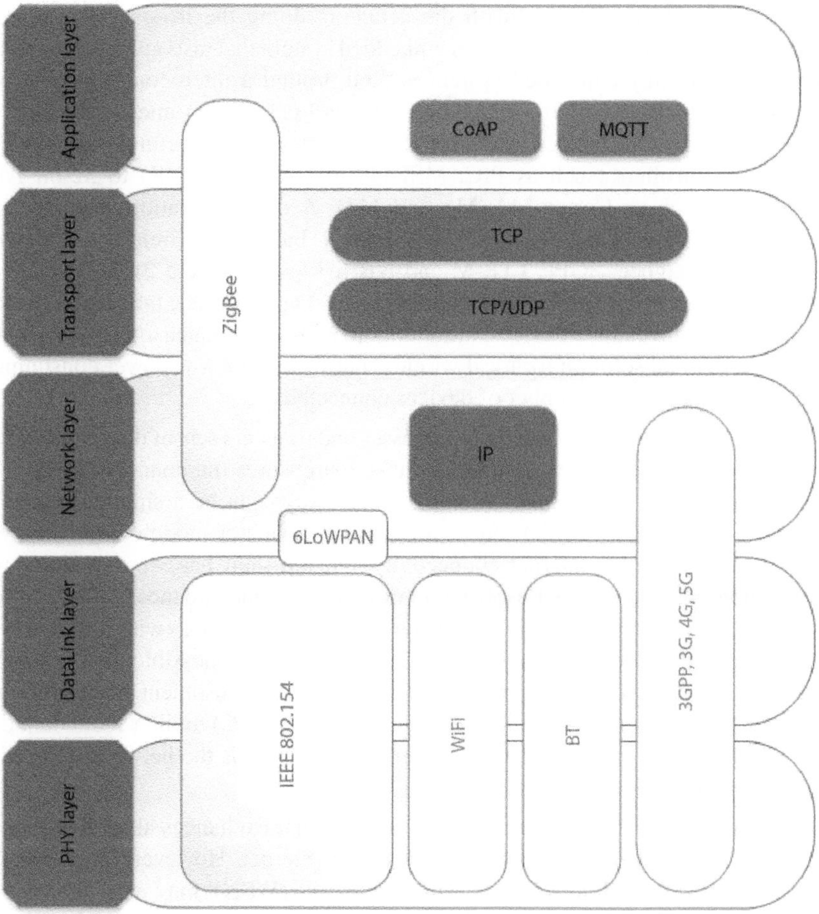

Figure 5.2. *Five-layer IoT and its equivalent OSI layers. For a color version of this figure, see www.iste.co.uk/ali-yahiya/tactile.zip*

The common layers between them are the perception, transport and application layers. The fundamental network infrastructure, as the name IoT suggests, is based on the IP protocol. Figure 5.2 shows the five-layer architecture of an IoT system. Note the similarity between the architecture and the open system interconnection (OSI). In order to realize an IoT system with this architecture, four main elements are needed. They are interconnected with intra- and inter-elements software and hardware architecture. Figure 5.3 shows the hardware architecture of the four main elements detailed below:

– **End devices** are physical hardware elements like sensors and actuators that collect data and perform actions. The function of sensors is to transform physical quantities into electrical ones, with the actuators doing the inverse. The sensors' electrical signals are acquired using an embedded system that also gives the actuators' commands. The tiny embedded system is built around a microcontroller that is at the heart of a hardware platform. Many cheap and performant microcontrollers are available in the market place, with different processor architectures ranging from 8-bit microcontrollers (such as PIC, MSP430 and ATMEL AVR) to 32-bit ARM architecture (such as Cortex-M0, M3 and M4). A communication protocol stack is built on the hardware platform. It is mainly based on wireless connectivity, such as WiFi, Zigbee, 3GPP, LTE-M and NB-IoT (see Figure 5.2). To prevent the heterogeneous protocol stacks at the end device level and facilitate the communication between different vendor's devices, the adoption of protocols, such as 6LoWPAN with IPv6, is an optimal solution for local wireless networks with low power consumption and the scalability of the number of devices connected.

– **Gateway** hardware collects, preprocesses and transfers sensor data from devices to the network, when a local wireless connection ensures the connectivity between the sensors and the Internet. An example of gateways can be a simple asymmetric digital subscriber line (ADSL) router, a Zigbee coordinator module interfaced with an IP stack module for internet connectivity. Off-the-shelf hardware modules with an operating system, such as Raspberry Pi or a Beagle bone, are the right choice for a gateway, since in specific IoT systems, heterogeneous end devices with different types of communication protocols may already exist. It is easily possible to add wireless hardware modules to this hardware for communication establishment. Moreover, these hardware modules can do considerable edge computing (EC) to filter redundant data, since reducing the redundant data before routing it towards the network will reduce the network load.

– The **network** is the communication medium that exchanges all of the data and commands. The backbone of the network is the Internet. However, other wired or wireless networks, such as wireless sensor networks (WSN), may exist between the end devices and the network. The cloud is an implicit part of the network, in which the server is used to work in a client–server combination, with the end devices and monitoring as clients. On the cloud, containers are now the common denominators for building cloud applications, and deploying them easily in different computing environments. Docker is used to package applications, with all of their dependencies, in a Docker image. The increased number of Docker containers makes the task of running and managing them difficult. The Kubernetes is the solution to orchestrating these containers.

– An **application** is a software component built on the end devices and the back-end. The application runs on a virtual cloud server. On the end device side, the application is running on the embedded processor to send or receive data according to a given lite protocol, such as CoAP or MQTT (see sections 5.5.1 and 5.5.2). The type of application varies between simple web-based data visualization dashboards and highly domain-specific mobile apps. Many application-layer protocols exist for IoT stacks, such as CoAP built on UDP. This is a lightweight protocol based on the *request/response* paradigm, which sends and receives the states of the systems. MQTT is another lightweight protocol that uses the *publish/subscribe* communication paradigm, and it is event-based. Both are used in M2M communications. Zigbee is a proprietary application layer protocol built on IEEE 802.15.4.

The software architecture ranges from simple software on a bare-metal-embedded device, to an OS-based device with APIs so that developers can connect the devices to the logical world easily (Taivalsaari and Mikkonen 2018). The software architecture is mainly required for the application layer's software since the other layers have their fixed functional architecture or middleware with API. Many development boards are available on the market with integrated development environments (IDE). They provide ease of use with fast market development.

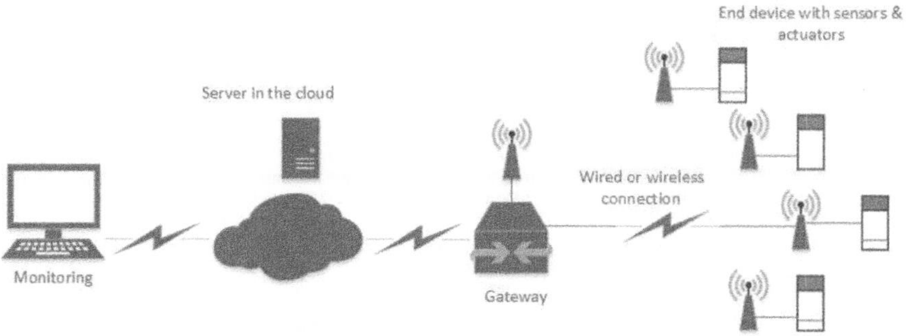

Figure 5.3. *Possible implementation of an IoT system with its main elements. For a color version of this figure, see www.iste.co.uk/ali-yahiya/tactile.zip*

5.5. IoT communication protocols

An IoT node is based on an embedded device that connects to the Internet and exchanges information with other devices over it. These devices have lightweight software to read the sensors built around the embedded device, a simple microcontroller. The devices should then adopt a way to exchange information. This is done by running lightweight protocols, such as MQTT and CoAP, in the application layer to encapsulate messages and data and add a small amount of overhead for establishing connections, providing a certain level of QoS. The packets are then passed through the transport, network and MAC layers of the Internet stack.

A detailed presentation of IoT communication protocols is given for each of the available IoT communication protocols in the following section.

5.5.1. *Message Queuing Telemetry Transport (MQTT)*

The MQTT protocol was initially designed to connect power-constrained devices over a low-bandwidth network. IBM engineers invented it in 1999. The protocol runs over TCP/IP or other similar network protocols. The protocol became open in 2010 and is currently standardized under the wing of OASIS (2019), where version 5.0 has been released.

The communication model is based on a central broker, where devices are connected to it as subscribers and publishers, as shown in Figure 5.4. Many subscribers (clients) and publishers can connect to the same broker. MQTT uses subject-based filtering for messages. Messages contain a subject that is used by the broker to determine whether a subscribing client gets the message or not.

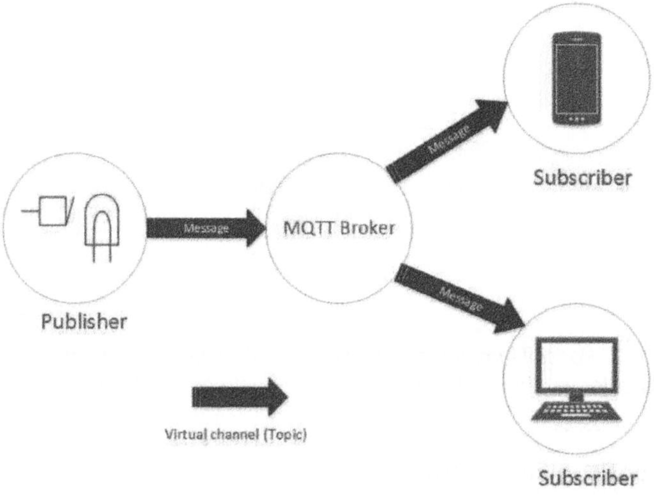

Figure 5.4. *MQTT publisher, broker and subscriber. For a color version of this figure, see www.iste.co.uk/ali-yahiya/tactile.zip*

It is a lightweight protocol consisting of a header with three parts, as shown in Figure 5.5. The first part is two bytes of a fixed header. It consists of a 1 byte control header and (1–4) bytes packet length. The second one is a variable header whose content depends on the packet type (message ID, topic name and client identifier). The third part is a payload where its presence depends on the packet type. For example, in a CONNECT packet, the payload contains the client ID, user name and password if they are present, while for a PUBLISH packet type, it is the message that is to be published.

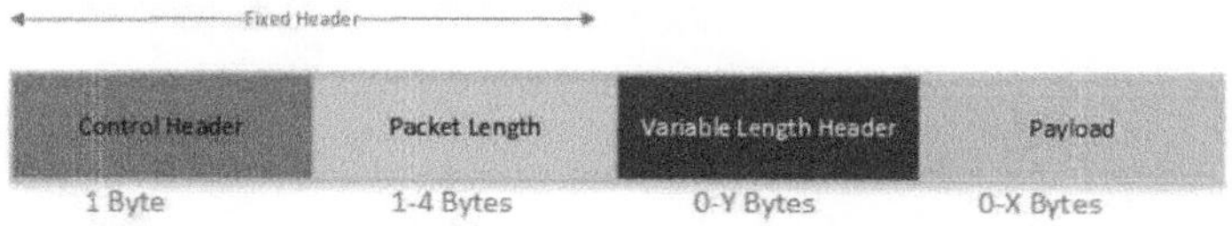

Figure 5.5. *MQTT standard packer structure. For a color version of this figure, see www.iste.co.uk/ali-yahiya/tactile.zip*

The MQTT has some essential features, such as the last will and testament, where the broker will send a message to all of the clients informing them that a client has disconnected ungracefully, by releasing a stored message containing information about the client to them.

A message delivery from a publisher client to a subscriber client happens in two hops: from the publisher to the broker and then from the broker to the subscriber, as shown in Figure 5.4. A similar handshaking procedure between publisher–broker and broker–subscriber follows. This procedure is determined by the required QoS of the message exchange between the publisher and subscriber.

Even though the MQTT is used on top of TCP/IP, data loss can still occur. Therefore, a QoS functionality is added to the protocol to reinforce the TCP/IP one. It is provided in three QoS levels: 0, 1 and 2, from low to high quality. The QoS 0 is equivalent to sending a message once by the publisher client, without expecting acknowledgment from the broker and the subscriber client. The message is then deleted and will not be transmitted again by the client as it is not stored. With QoS 1, it is guaranteed that the message will be delivered at least once, possibly more than once. This is obtained first by keeping the publisher's message and checking that the acknowledgment signal PUBACK returned from the broker. If the message is not received, retransmission will occur. In case the duplication of the received message is harmful, another level of quality, QoS 2, is available. It guarantees that there is only one delivery, and it will not be repeated. This happens in a two-step acknowledgment procedure, as shown in Figure 5.6.

Data security issues in the MQTT are handled by the transport layer security protocol (TLS) at the TCP layer, where the data is encrypted to guarantee its integrity after transmission. The MQTT can also add another security level at the broker level, where a client ID, user name, password and client certificate allow authentication.

Open-source implementations of the different parts of the protocol are available. The client and broker of the protocol are available. The client, for example, is implemented by the Eclipse Paho. The broker is implemented by the Mosquito on Eclipse (Kayal and Perros 2017).

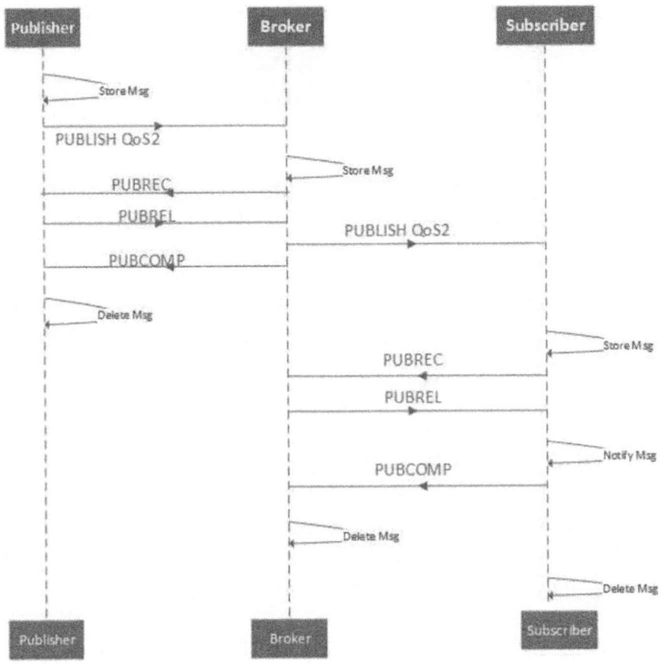

Figure 5.6. *MQTT quality of service 2. For a color version of this figure, see www.iste.co.uk/ali-yahiya/tactile.zip*

5.5.2. *Constrained Application Protocol (CoAP)*

The CoAP is an asynchronous web transfer protocol, optimized for constrained nodes in IoT and M2M network applications. It is designed for low-power and lossy wireless networks. The standard was proposed by IETF in 2014 (Internet Engineering Task Force 2014). Like HTTP, it uses the REST model and RESTFul API, which is used in web services. It is mainly used in low-power wireless IoT applications, with an IPv6 network layer on a simple 8-bit microcontroller. The "constrained" adjective is added because of the low-power wireless network used with the IPv6 network layer, such as 6LoWPAN, and also due to the devices that need constrained resources.

The structure of the message header is shown in Figure 5.7, where (Ver) is a 2-bit field for determining the version of CoAP, T is a 2-bit field for determining the type of message, for example CON (confirmable), NON (non-confirmable), ACK token length, and TKL, which is a 4-bit field for the indication token field. The code field specifies whether the message is a request. The message ID field is used for the detection of duplicate messages (acknowledgment) or RST (reset), of variable length of response or empty. Figure 2.14 shows the structure of the message header.

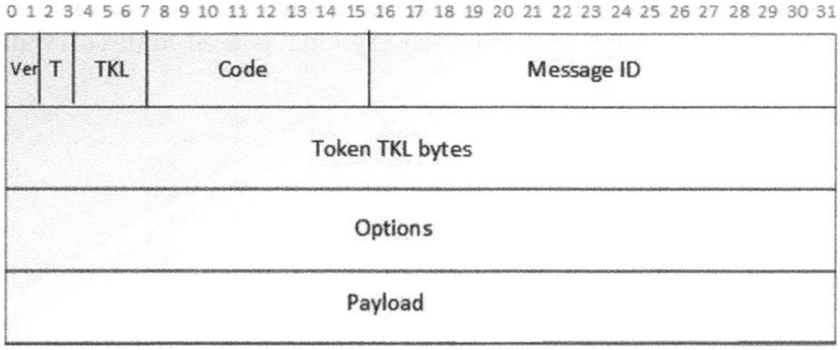

Figure 5.7. *The structure of the CoAP protocol message header*

The architecture of participating entities in CoAP is a client–server one. The server makes resources available under a URL, and the clients access these resources using methods such as GET, PUT, POST and DELETE.

A useful feature of CoAP is the possibility of sending confirmable and unconfirmable messages. When the receiver gets a confirmable message, it will return an acknowledgment. If the sender did not receive the acknowledgment, it will continue to send until a timeout is attained.

The CoAP runs over UDP and supports the use of multicast IP destination addresses. This is useful in IoT applications where a sensor wants to send a reading to multiple endpoints. In the multicast case, to prevent the congestion of the network, a non-confirmable message is sent. However, if a request needs a response by the server, it should not respond immediately, but after a period of leisure.

The architecture in Figure 5.8 shows how the constrained network is connected to the Internet with or without a proxy. The CoAP protocol can either be mapped to the HTTP through a proxy or establish a direct connection between a resource providing the server on the Internet and the request originating client. The nodes (c) are connected in the constrained network using CoAP, where they can function as sender and recipient endpoints.

Concerning security, as mentioned in the case of MQTT, the three elements authentication, encryption and integrity are guaranteed by using datagram transport layer security (DTLS) over UDP (see Datagram Transport Layer Security Version 1.2 by Internet Engineering Task Force (2012)).

Web-of-Things, in analogy to the IoT, uses web browsers to communicate with things by applying the RESTful paradigm (Richardson *et al.* 2007; Mingozzi *et al.*

2014) to constrained devices. The CoAP is designed for this case since it is a version of HTTP with a reduced overhead and limited complexity, which fits the limited capabilities of constrained devices. As shown in Figure 5.8, the clients can ask for services from the nodes, using the REST interface in the same way as when accessing web services. For example, resource discovery GET is used to discover all of the devices.

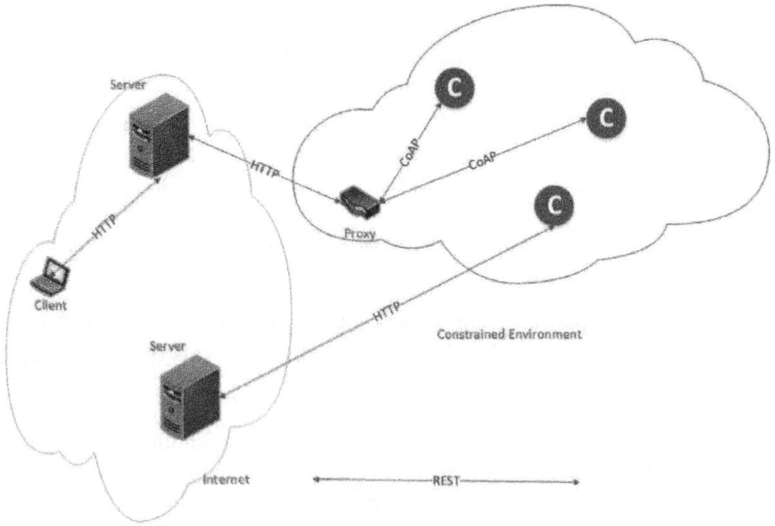

Figure 5.8. *CoAP architecture. For a color version of this figure, see www.iste.co.uk/ali-yahiya/tactile.zip*

5.5.3. *Data Distribution Service for real-time systems (DDS)*

The Object Management Group (OMG) DDS standard is the first open international standard directly addressing publish-subscribe middleware for real-time systems. Its most remarkable features are the real-timeliness, reliability and delivery deadlines. Hence, it is convenient for distributed systems. Other advantages of this protocol are its dynamic scalability and its support for one-to-one, one-to-many and many-to-one and many-to-many communications. The DDS application middleware can be compiled for any platform thanks to the platform-specific mapping of interface definition language (IDL), which is a descriptive language used to define data types and interfaces independently of the programming language or operating system/processor platform.

DDS is a data-centric protocol where publisher nodes transmit the state (data) of their world and update it if an attribute changes. The communication infrastructure will maintain the status. The subscribers interested in the data, also called "topic", will subscribe to the topic to be updated. Fields of a given topic "key" will provide detailed specifications of the data. This is in contrast to a message-centric system, where commands are sent to change the world's state, which will be reconstructed by an application on the remote client node. The difference between data-centric and message-centric is demonstrated in Figure 5.9, where the client node application memorizes the states in the message-centric system. The messages (commands c_is) coming from another node will modify the state in the destination client addressed by the transport and network layers. In other words, the messages are verbs of action to the destination clients. In data-centric, the transport layer is aware of the destination as a client subscribes to a data name and type. The data is then analogous to "nouns" compared to the messages equivalent to "verbs" in languages.

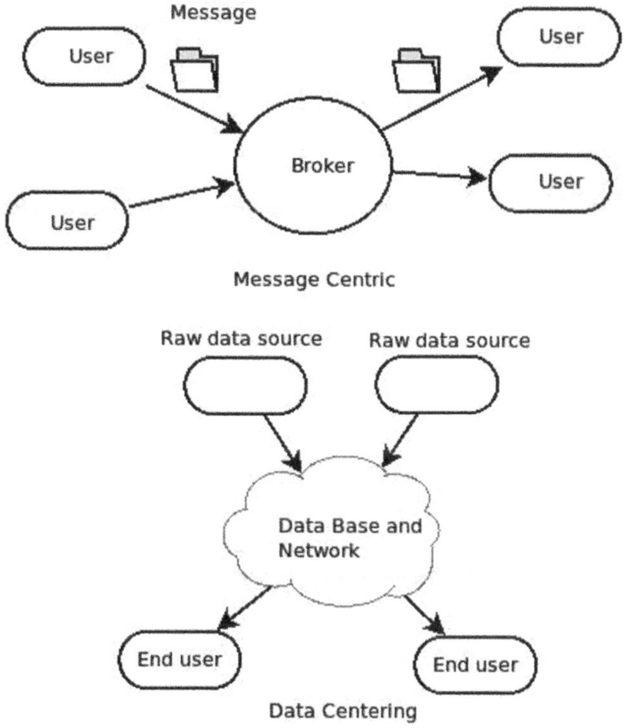

Figure 5.9. *Data-centric versus message-centric*

As shown in Figure 5.10, the DDS architecture consists of DomainParticipant, Publisher, Subscriber, DataWriter, Topic and DataReader entities. The DomainParticipant is the top-level entity in the architecture. It is created by an application and is used to join a DDS domain, where a DDS domain is a logical network of applications that can communicate with each other. In other words, a

DomainParticipant is a container for entity objects that all belong to the same DDS domain. The DomainParticipant creates other entities such as the DataWriter and DataReader. The former writes typed data on a topic. An aggregation of DataWriter objects makes a publisher that disseminates information. The DataReader reads typed data from a topic, an assembly of them is responsible for receiving data. Concerning the bus system, DDS uses UDP by default, but it can also support TCP. An interoperability problem may result from different transport protocols used by the vendors in their implementation of the DDS. To overcome this problem, the real-time publish subscribe (RTPS) (Object Management Group 2018b) wire protocol is used. It represents the DDS interoperability protocol that allows data sharing among different vendor implementations.

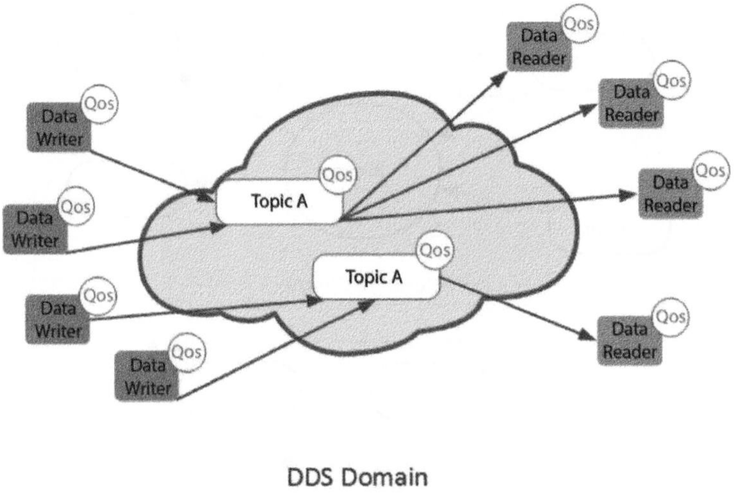

Figure 5.10. *Architecture of a DDS protocol to connect applications systems. For a color version of this figure, see www.iste.co.uk/ali-yahiya/tactile.zip*

Compared with the MQTT, the DDS does not depend on a broker and different clients can communicate in a peer-to-peer manner.

DDS's QoS policy is such that each entity has an associated QoS comprised of a list of policies, depending on the entity type (see Comprehensive Summary of QoS Policies (Cheat Sheet) by Real-Time Innovations, Inc. (2013)). It configures the different aspects of the entity's behavior. For example, a DataWriter on a Publisher can offer a QoS contract where it commits to provide data each deadline period. A DataReader on a Subscriber requires a QoS contract that expects data reading each deadline period. A publisher can also "promise" whether it will use a best-effort UDP-like or reliable TCP-like protocol for transport. A subscriber can tell the middleware that it only wants to see every third topic sample and/or only the samples in which the specific data-field-filtering criteria are met.

Regarding the protocol's security model, DDS has a service plugin interface (SPI) architecture (Object Management Group 2018a) at the middleware level. This allows users to customize authentication, control access, encrypt, authenticate messages, sign digitally, and log and tag data.

In Yuang Chen (2016), a quantitative comparison of latency, bandwidth and packet loss point of views is done. DDS shows good reliability and latency performance with respect to similar TCP publisher/subscriber protocols like MQTT. The price to pay for this is the increased bandwidth used by DDS. For more detailed references, the reader can refer to de C Silva *et al.* (2019), where a comparison with other IoT protocols is given.

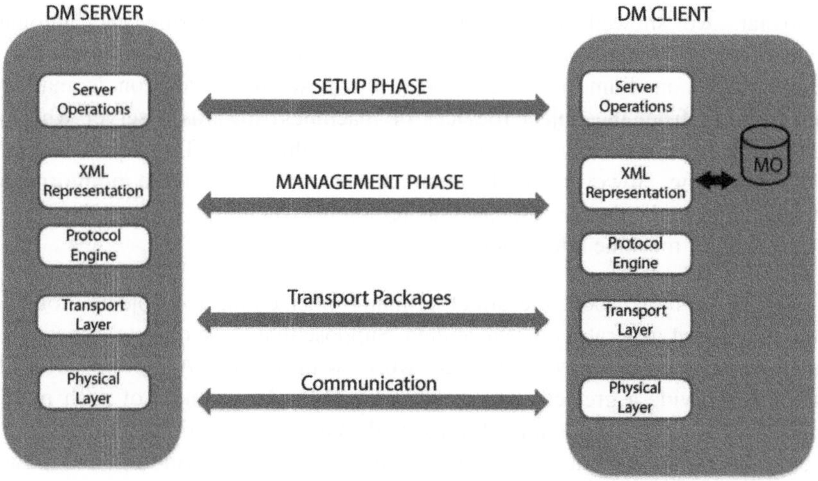

Figure 5.11. *Architecture of the OMA-DM protocol. For a color version of this figure, see www.iste.co.uk/ali-yahiya/tactile.zip*

5.5.4. *Open Mobile Alliance Device Management (OMA-DM)*

This protocol allows management commands to be executed on nodes. It is based on where a DM server sends secure management commands to a client and a DM session consists of a request/response transaction (see OMA Device Management Protocol by OMA (2016)). The server manages the client nodes in two ways, either by reading and setting parameter keys and values or by installing, upgrading or uninstalling software elements when running software on the client side.

A session consists of two phases: *setup* and *management*. In the former, authentication and device information is exchanged. In the latter, the server will issue commands to be processed by the client. The communication messages and data use the XML format.

Figure 5.11 shows the architecture of the protocol. The server and client connection is either a wired (USB, RS-232 or Ethernet) or wireless physical layer (GSM, CDMA or Bluetooth). Regarding the transport layer, it is the same as in HTTP.

The security level of this protocol is not high. It is only based on authentication by sending credentials from the client, which will get authorization from the server. The credentials are hashed using MD5 crystallographic hashing.

5.6. Internet of Everything (IoE)

So far, we showed that the IoT is mainly a monitoring and automated data collection system, using sensors, embedded systems and Internet as the communication medium. In other words, the flow of information is mainly in one direction, from the object (devices or machines) towards a server where the information is processed and monitored, meaning that the IoT is a passive system in its functioning. However, simple commands can be sent from the monitoring side to take action on the device side, such as remotely switching a device ON and OFF, using human or machine interaction.

Now let us think about the effect of many other connected objects around the initial object that generates correlated data. Suppose that these objects' functions are related by determining how each object works with the rest. A process adopts this relation to provide a greater value to the ensemble. An example of such multiple connected objects is an automated home with many connected devices, such as an air conditioner, an oven, a coffee machine, a TV, ventilation windows and a garage door. These objects will generate heterogeneous data, which may be incompatible, correlated and/or independent. A process is an algorithm that manages objects' data, according to their relationship and the data they provide. After data manipulation, the process delivers the right information to the right machine or person at the right time. People or business managers can interact with this process through the Internet for its configuration. This whole system, composed of things, networks, people, data and processes, makes the Internet of Everything system.

The IoE system uses hyper-connectivity between smart objects generating data, people connected through a social media network and algorithms that deal with the considerable amount of data generated by the smart objects and people. Figure 5.12 shows the different constituents of an IoE system, as suggested by Cisco. Therefore, other technologies are necessary for IoE in contrast to the simple connectivity of the IoT. In the following section, we explain these enabling technologies.

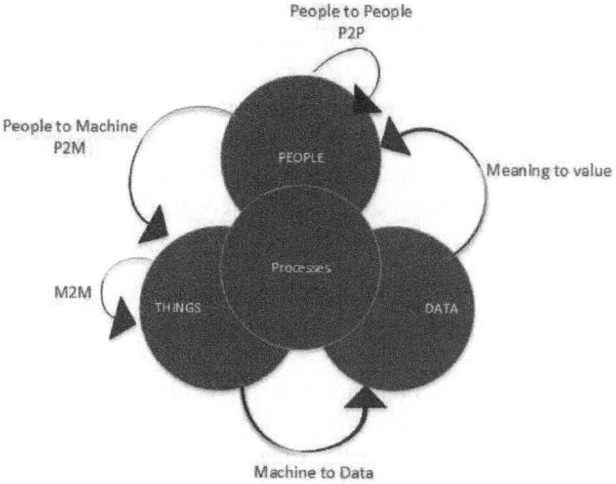

Figure 5.12. *People, process, things and data interactions. For a color version of this figure, see www.iste.co.uk/ali-yahiya/tactile.zip*

5.6.1. *Enabling technologies for the IoE*

Technology is the basic driver for the emergence of IoE. The enabling technologies, mainly related to the smartness of the objects listed by Langley *et al.* (2021), are:

– interconnectivity: the interconnection of many devices, with their smart ability in sensing and reacting to the environment, helps in increasing the smartness of the IoE, by exchanging quality data through a network of specialized smart components. Computational power for the interconnected devices can be added through either edge computing or cloud services;

– big data: improved algorithms to interpret the increasing amount of quality data is another key enabler of IoE. Thanks to distributed systems, it is now possible to deal with significant volumes of data near the source, without sending it through the network. This local, close to the source, analysis will endow the devices with an adaptive optimized behavior, such as energy efficiency;

– artificial intelligence (AI): the connected devices should be smart and autonomous in sensing reasoning, making decisions and performing actions. These capabilities are provided by AI systems and algorithms;

– semantic interoperability: heterogeneous connected devices should understand and exchange information, data and knowledge in a meaningful way, in the same manner as the semantic web, which allows heterogeneous data in web pages to be structured, linked and retrieved. The development of successful interoperability between devices can be how things and systems expose data and metadata through API. In this case, the devices should share information concerning these APIs.

The range of IoE applications will depend on the smartness of the things (Langley *et al.* 2021). The higher level of smartness of things will increase their collaborative capabilities, which will impact the business models by opening up new opportunities. Also, increased smartness moves the system connectivity from closed to open interoperable systems.

5.7. Protocol comparisons and the readiness for TI

Previously, we listed the communication protocols used in IoT systems. Here, we compare them and investigate the possibility of them being used in a TI system. Table 5.2 summarizes their comparison using different aspects: applications, QoS, data dissemination and interoperability.

The comparison shows that DDS conforms to the TI requirements the most across many aspects. The most important one is real-timeliness, which is critical in the case of TI systems. No work has been found applying the DDS protocol for data exchange between master and slave devices. It is mainly because in TI applications, only a single master is used. The DDS could be a solution for a collaborative TI application where many masters can intervene. For more insight, Yang *et al.* (2012) used it in industrial applications, with a detailed analysis of the latency between nodes.

	DDS	MQTT	CoAP	OMA-DM
Architecture	P2P	Broker based	Client/server	Client/server
Transport	UDP/TCP/RTPS	TCP	UDP	localUSB and HTTP
Data dissemination	Unicast, multicast	Unicast, multicast	Unicast	Unicast
QoS	20 criteria	four levels	Message acknowledgment based	Not available
Interoperability	Different platforms and programming languages	Different platforms and programming languages	different platforms and programming languages	Different platforms and programming languages
Real-timeliness	yes	no	no	no

Table 5.2. *Comparison IoT protocols*

5.8. TI-IoT models and challenges

Following the brief review of the IoT architecture and protocols, a comparison between the IoT and TI technology is necessary. A reference architecture for TI is suggested in the IEEE P1918.1 standards, as shown in Figure 5.13. Comparing it to the one for IoT, given in Figure 5.1, reveals many similarities. However, a more in-depth analysis will show the main differences as follows:

– the devices: in IoT, the devices used can be, in addition to sensors/actuators, data-carrying devices connected to the physical device to send and receive data for automation and monitoring purposes;

– the connectivity: the connectivity of devices to the network can be wired or wireless. Since the TI's essential requirement is the ultra-low latency and availability, this restricts the connectivity possibility to 5G networks. While most IoT services are not real-time, connectivity through Ethernet connection, WiFi, NB-IoT and LTE-M is possible;

– data processing and communication protocol: in the TI, the data is streaming continuously in real time, while in IoT, the information is mostly in a burst that can be aggregated, processed and stored, before monitoring or utilization for automation purposes;

– cloud: in the IoT, the cloud is used for IaaS (Infrastructure as a Service) or PaaS (Platform as a Service), but also for data processing. The cloud is not available in the TI since the data should be conveyed by the network in real time. The communication network is then called the network domain;

– the applications and services exist in both the IoT and TI to provide the required services, using the application layer to collect and preprocess data from sensors in order to decode and direct received data to actuators.

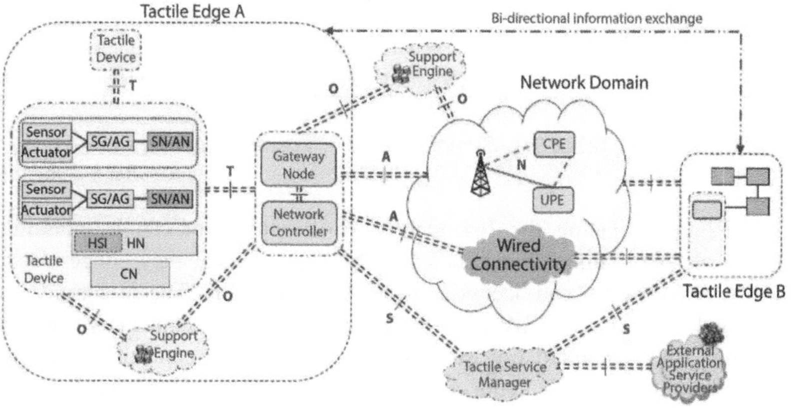

Figure 5.13. *IEEE P1918 Tactile Internet reference architecture (Holland* et al. *2019). For a color version of this figure, see www.iste.co.uk/ali-yahiya/tactile.zip*

A summary of the comparison is given in the Table 5.3, from different requirement aspects.

Requirement	IoT	Tactile Internet
Real-time	Non-real-time	Real-time with reaction time (1–10 ms)
Function	Monitoring, data processing and control	Control with feedback †
Connectivity	Internet, wired and wireless	5G
Data processing	Aggregation, analytic	Haptic data compression and modeling
Architecture	Three- to five- layers (devices, edge, network, cloud, application)	Master domain, network domain, teleoperated domain
Applications	(Non-critical) agriculture, smart city	(Critical) health care, traffic, robotics and manufacturing
Scalability	Large-scale deployment	Near-future large-scale deployment
Network and interfaces	Simple availability and reliability	High availability and reliability
Control process	Automated	Interactive
E2E communication model	Client–server	Master–slave
Power consumption	Restricted power consumption	No restrictions
Network configuration	Centralized around the server	End-to-end connection
Human interaction	Human in the loop to monitor and manage remotely	Human in the loop to control remotely (human-to-robot)

Table 5.3. *Summary of IoT-TI comparison*

The challenges facing the IoT technology development can be classified as technological, environmental and societal. More specifically, they concern security, scalability, connectivity and access control of IoT systems. The increased number of connected devices will have negative environmental consequences from energy consumption and generated e-waste. This may be controversial because the IoT technology optimizes and reduces power consumption when used for home automation, smart city and energy grid control, and also, when used in agricultural applications to optimize water consumption. Hence, IoT is an opportunity and a challenge (Radu 2018). On the societal level, the increased services provided by the

IoT technology, such as surveillance cameras, are beneficial in reducing the number of crimes. However, they can cause some concerns among civil rights advocates.

The main challenges facing TI implementation are the ultra-low latency of round trip of a data packet, high reliability, availability of connection and safety and security of the TI application. The available communication technology to support the above challenges is 5G.

To sum it up, IoT and TI, with the advent of 5G, share many similarities and some subtle differences. The difference is clarified in the mode of human interaction in the case of TI, which is human-centric, and the machine is complementing and augmenting human capabilities (Maier *et al.* 2016).

5.9. Edge computing in the IoT

The deluge of data generated from the proliferation of connected objects to the Internet on one hand, and the convergence of IoT and cloud techniques on the other, will require cloud processing power and storage. This may increase the load on the network to unacceptable ranges. An efficient solution to this problem is to process at the cloud's edge, close to the data source. Edge or fog computing is added to the IoT architecture to decrease this load, by offering some processing, like data analysis, closer to the source. In a use case, such as a smart city, global handling of the massive amount of data generated by the different city services is needed, rather than a siloed processing of other city areas: public lighting management, health care system, car parking and public transportation. The transmission of the increased amount of data over the network will result in congestion with a degraded QoS, due to the increased transmission delays and the decreased throughput. For example, a delayed health care emergency has a severe impact.

Historically, edge computing can be traced as follows (Khan *et al.* 2019): the computer systems evolved from the mainframe, where all of the processing power is confined in, to independent smaller desktops. However, the desktop is attained in computing power and memory space. The earlier solution was to connect them to a server system located closer to them in a client–server manner, in order to serve many desktops simultaneously. The growth of the number of software applications to increase productivity added more expenses to users, therefore, the cost of license rights, on one hand, and the cost of infrastructure and platforms, on the other hand, needed to be reduced. The computing power has again been shifted towards the network to establish a cloud system of distributed computing and memory platforms across the Web. Later, with the advent of mobile computing devices and their capability to connect to the network ubiquitously, the need appeared for fixed computing devices within the core network to support them in a cloud system. Finally, the proliferation of the IoT devices and the increased demand for computing power required the extension of the cloud computing resources towards the end devices (edge computing), in order to decrease the cloud network congestion with unuseful flow to reduce the latency. Figure 5.14 shows how the computing power shifts to the edge through three different configurations, which will be explained in the next sections.

The main aim of concentrating the computational power is to reduce the expensive computing hardware and memory facilities; this attributes features such as, resource pooling and storage capacity to the small IoT devices connected to the network.

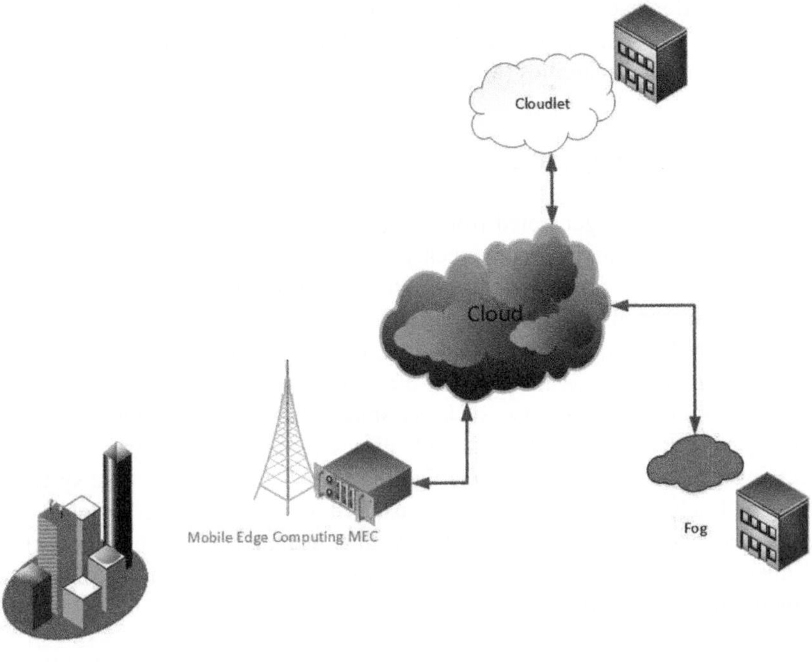

Figure 5.14. *Edge computing in IoT systems. For a color version of this figure, see www.iste.co.uk/ali-yahiya/tactile.zip*

5.9.1. *Edge computing paradigms*

Many projects in academia and industry supported edge computing, such as cloudlet, fog, mobile edge computing (MEC), Nebula and FemtoCloud, among others (Pan and McElhannon 2018), to implement the edge computing paradigm. Moreover, software-defined networks (SDN) and virtualized network functions (VNF) combined with the cloud edge platforms will virtualize the network, automatically configure IoT devices connected to the network and add new virtual functionalities, as shown in Figure 5.15. The virtualization of the network is useful in providing reliable connectivity. In contrast, the functionalities' virtualization is useful for changing or adding new services and functionalities to the network. Both VNF and SDN solve the problems of heterogeneity, interoperability and scalability of IoT devices.

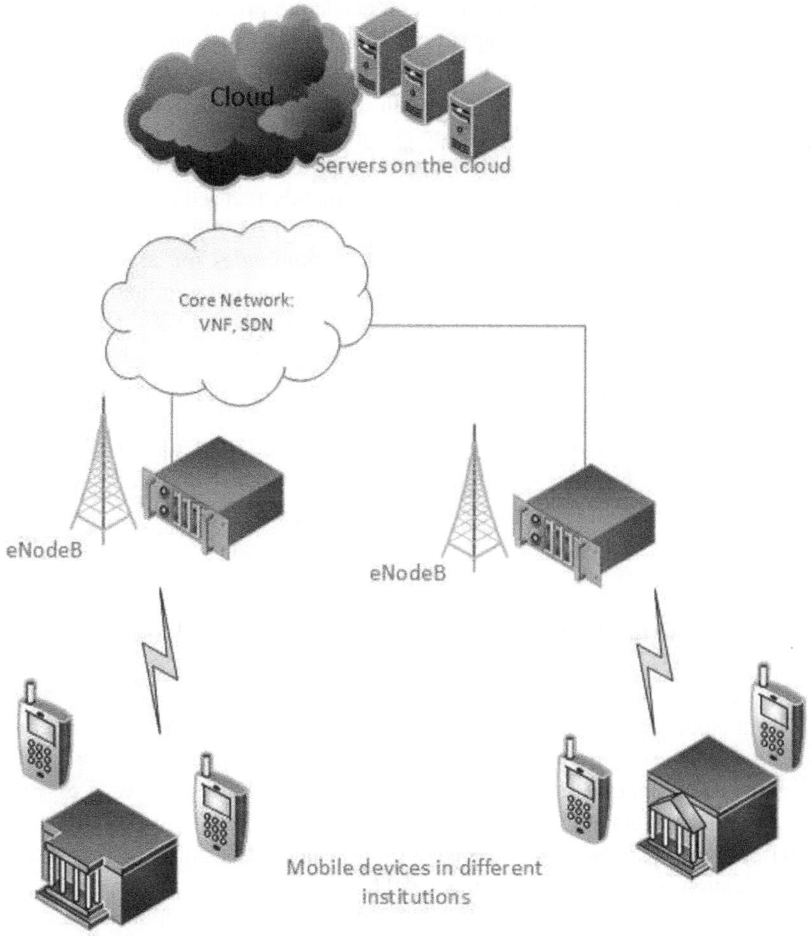

Figure 5.15. *SDN and VNF in the core network and the MEC access to them. For a color version of this figure, see www.iste.co.uk/ali-yahiya/tactile.zip*

Cloudlet is based on a multi-tier architecture, where the Cloudlet is a computing device or a "data center inside a box" located between the Internet and the mobile units. It is close to the end-users' wireless mobile or wired device, generally any "thin client" device. Inside the box, virtual machines (VMs) are instantiated and allocated dynamically according to the end devices requests for assistance. Similar to Cloudlet, (Bonomi *et al.* 2012) it aims to provide compute, storage and services in the proximity of the end-users, as well as to reduce network congestion, E2E latency and enhanced security levels by reducing the core environment operations. However, sophisticated security algorithms should be embedded on the edge devices. The difference between these two technologies is that the first is an academic project and the latter is from the industry. The mobile-edge computing (MEC) initiative from ETSI (2016) is to prepare the transition towards 5G and the IoT. It aims to provide a new ecosystem and value chain for different actors: application developers, content providers, network operators and customers. The MEC servers are deployed at the base stations, such as LTE (eNodeB) and 3G Radio Network Controllers (RNC). The main difference between Cloudlet, Fog and MEC is that MEC is mainly focused on cellular networks.

To summarize, the SDN and NVF enable technologies for edge computing to easily access the edge computing devices for their configuration and management. While these technologies are developing, Cloudlet, Fog and MEC architectures are suggested to implement the early edge computing systems in cloud-based networks. The MEC and Fog architecture can be more adapted to IoT applications.

A detailed survey on mobile edge computing is given in Mao *et al.* (2017). A use case example of heavy edge computing is provided in Li *et al.* (2018), where deep learning is introduced at the edge to analyze the offloaded data from the end devices (nodes).

5.10. Real-time IoT and analytics versus real time in TI

The pervasive characteristic of the IoT leads to huge data from several smart connected objects. On the other hand, humans are part of the IoT system interaction, since it collects users' information. Therefore, a real-time analysis of the amount of automated data generated from heterogeneous and distributed sources is essential to extract useful information for humans. Nevertheless, the real-time analytics required for IoT systems is not as critical as in TI. The TI criticality is inherent since real-time communication is required for a kinesthetic return signal from a teleoperated system to the human operator (Haddadin *et al.* 2019). However, in IoT applications, real time stands for a fast response, without a deadline constraint to respect.

5.11. From IoT towards TI

Following the connection of millions of computers using the Internet and millions of mobile phones via the wireless network to the Internet, billions of objects are now connected via the Internet and wireless networks, merging the physical world and the virtual world. Up until the advent of IoT, the Internet provided a connection from anywhere, at any time, for anything and anyone, for any service and on any network (Vermesan *et al.* 2009). Currently, with the advent of 5G, any haptic interaction experience is realized, where it provides the infrastructure and medium to establish haptic communications. The essential requirement for any haptic interaction dimension is low latency transmission. This additional dimension is a leap in today's IoT technology for the Tactile Internet, which delivers physical, tactile experience remotely. Figure 5.16 shows that IoT and TI may share the mission-critical communication that can be provided by 5G, which can be useful for both.

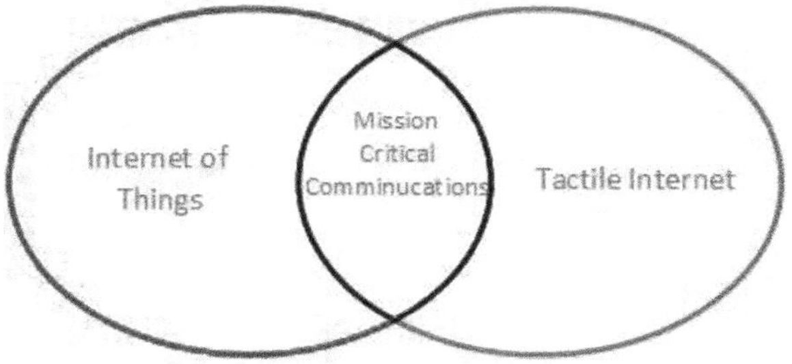

Figure 5.16. *Mission-critical communication for IoT and TI (Zhang and Fitzek 2015). For a color version of this figure, see www.iste.co.uk/ali-yahiya/tactile.zip*

As in the case of the IoT, some enabling technologies are RFID, wireless networks and sensors, and embedded system miniaturization. In the case of the TI, ultra-low latency reliable connection is the enabling technology. The range of applications between IoT and TI is different, as shown in Table 5.3, but both are built around sensors and actuator devices and network connection. Figure 5.17 shows the similarities and differences between the IoT and TI, where the main difference to be noted is the ULLRC constraint on the haptic communications through the network and the human in the loop, since the interaction with the remote physical world is through tactile perception.

The most appealing application, among others, of TI would be the teleoperation of robots. In such applications, a complex system, such as a robotic arm, is remotely controlled, with haptic feedback to the master controller to improve the remote arm control, while it evolves in the environment. The haptic sensors supply details of the object that the user is interacting with in real time. The next chapter tackles this remote robotic control via a network system to provide the required Quality of Experience (QoE) and the quality of the executed task on the slave side. Haptics will play a significant role in the future Internet, which can be the critical point of entry into a full-sensory virtual reality. It will create augmented reality and the sensation of immersion. The capacity to uniquely address most objects of our real world, with the help of haptics, makes it possible to create an augmented world that will increasingly resemble our physical world and reduce the gap between the digital and physical worlds.

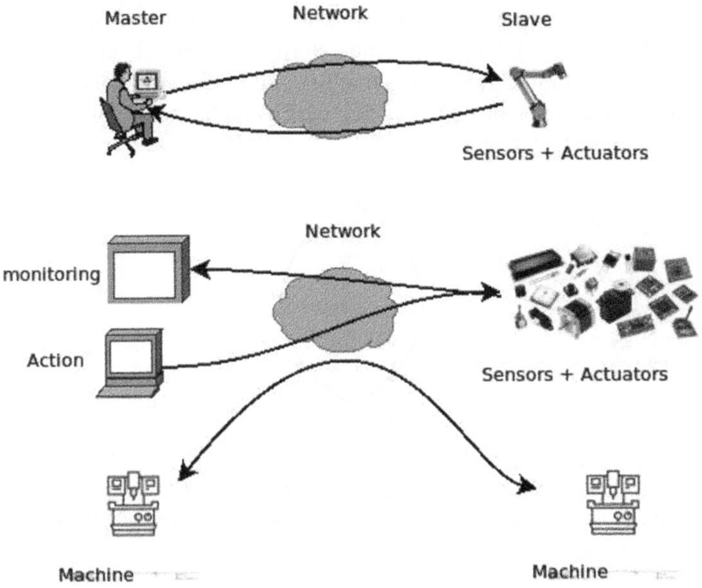

Figure 5.17. *Communality and difference between the IoT (bottom) and the TI (top). For a color version of this figure, see www.iste.co.uk/ali-yahiya/tactile.zip*

5.12. Conclusion

This chapter covered the IoT architecture and the communication protocols of IoT. It presented the connection of objects to the Internet and M2M. The aim was to follow the evolution of the connected things' paradigm to the tactile experience paradigm, focusing on the communication protocols, specifications and applications. We were reminded of the three- to five-layer architecture of IoT. An edge device is an interface that can extend between the end device and the cloud. A local network can also exist between the end device and the edge closer to the device. This layer provides connection and communication operations, in addition to some computational functions. The cloud centralizes the whole system. Service applications are offered, such as data analytics, monitoring and centralized device management. It is reminded that connectivity, edge capabilities and security are the main challenges faced in IoT systems and their communication protocols.

The Internet of Everything expands the IoT concept by including data, people, processes, and network connectivity.

Regarding the TI, a standard three-layer architecture consisting of the master domain, network domain and slave domain is the dominant one. Most of the computations are held in the slave and master domains, while the network domain provides a reliable low latency connection between the master and slave domains. The latter two properties of the network domain are also the main challenges for a TI operation, especially for critical systems. TI is not next generation IoT. However, a new paradigm, such as the Tactile IoT can emerge in the future, when the physical world is merged with the virtual and digital worlds and then is again transformed into the physical world at the other end of the connection.

5.13. References

Acatech. (2011). *Cyber-Physical Systems*. Springer-Verlag, Berlin.

Bonomi, F., Milito, R., Zhu, J., Addepalli, S. (2012). Fog computing and its role in the internet of things. *Proceedings of the First Edition of the MCC Workshop on Mobile Cloud Computing, MCC 12*. Association for Computing Machinery, New York, 13–16 [Online]. Available at: https://doi.org/10.1145/2342509.2342513.

de C. Silva, J., Rodrigues, J.J.P.C., Al-Muhtadi, J., Rabêlo, R.A.L., Furtado, V. (2019). Management platforms and protocols for internet of things: A survey. *Sensors*, 19(3), 676.

ETSI (2016). Mobile Edge Computing (MEC); Framework and Reference Architecture. ETSI GS MEC 003 V1.1.1.

Gardašević, G., Veletić, M., Maletić, N., Vasiljević, D., Radusinović, I., Tomović, S., Radonjić, M. (2017). The IoT architectural framework, design issues and application domains. *Wireless Personal Communications*, 92, 127–148.

GrØnbæk, I. (2008). Architecture for the Internet of Things (IoT): API and Interconnect. *Second International Conference on Sensor Technologies and Applications (SENSORCOMM 2008)*, 802–807.

Haddadin, S., Johannsmeier, L., Díaz Ledezma, F. (2019). Tactile robots as a central embodiment of the tactile internet. *Proceedings of the IEEE*, 107(2), 471–487.

Holland, O., Steinbach, E., Prasad, R.V., Liu, Q., Dawy, Z., Aijaz, A., Pappas, N., Chandra, K., Rao, V.S., Oteafy, S., Eid, M., Luden, M., Bhardwaj, A., Liu, X., Sachs, J., Araúo, J. (2019). The IEEE 1918.1 "Tactile Internet" standards working group and its standards. *Proceedings of the IEEE*, 107(2), 256–279.

Internet Engineering Task Force (2014). The Constrained Application Protocol (CoAP) [Online]. Available at: https://tools.ietf.org/html/rfc7252.

Internet Engineering Task Force (2012). Datagram Transport Layer Security Version 1.2 [Online]. Available at: https://tools.ietf.org/html/rfc6347.

Jamali, M.A.J., Bahrami, B., Heidari, A., Allahverdizadeh, P., Norouzi, F. (eds) (2020). IoT architecture. *Towards the Internet of Things Architecture, Security and Applications*. Springer.

Kayal, P. and Perros, H. (2017). A comparison of IoT application layer protocols through a smart parking implementation. *20th Conference on Innovations in Clouds, Internet and Networks (ICIN)*, 331–336.

Khan, L.U., Yaqoob, I., Tran, N.H., Kazmi, S.M.A., Dang, T.N., Hong, C.S. (2019). Edge computing enabled smart cities: A comprehensive survey [Online]. Available at: http://arxiv.org/abs/1909.08747v1.

Langley, D.J., van Doorn, J., Ng, I.C., Stieglitz, S., Lazovik, A., Boonstra, A. (2021). The internet of everything: Smart things and their impact on business models. *Journal of Business Research*, 122, 853–863 [Online]. Available at: https://www.sciencedirect.com/science/article/pii/S014829631930801X.

Laszlo, M. (2014). Cyber-physical production systems: Roots, expectations and R&D challenges. *Procedia CIRP*, 17, 9–13.

Li, H., Ota, K., Dong, M. (2018). Learning IoT in edge: Deep learning for the Internet of Things with edge computing. *IEEE Network*, 32(1), 96–101.

Maier, M., Chowdhury, M., Rimal, B.P., Van, D.P. (2016). The Tactile Internet: Vision, recent progress, and open challenges. *IEEE Communications Magazine*, 54(5), 138–145.

Mao, Y., You, C., Zhang, J., Huang, K., Letaief, K.B. (2017). A survey on mobile edge computing: The communication perspective. *IEEE Communications Surveys Tutorials*, 19(4), 2322–2358.

Mingozzi, E., Tanganelli, G., Vallati, C. (2014). Coap proxy virtualization for the web of things. *IEEE 6th International Conference on Cloud Computing Technology and Science*, 577–582.

Pan, J. and McElhannon, J. (2018). Future edge cloud and edge computing for internet of things applications. *IEEE Internet of Things Journal*, 5(1), 439–449.

OASIS (2019). MQTT Version 5.0 [Online]. Available at: https://docs.oasis-open.org/mqtt/mqtt/v5.0/mqtt-v5.0.pdf [Accessed January 2021].

Object Management Group (2018a). DDS Security [Online]. Available at: https://www.omg.org/spec/DDS-SECURITY/About-DDS-SECURITY/ [Accessed January 2021].

Object Management Group (2018b). The Real-time Publish-SubscribeProtocol (RTPS) DDS Interoperability Wire Protocol Specification [Online]. Available at: https://www.omg.org/spec/DDSI-RTPS/2.3/Beta1/PDF.

OMA (2016). OMA Device Management Protocol [Online]. Available at: http://www.openmobilealliance.org/release/DM/V1_3-20160524-A/OMA-TS-DM_Protocol-V1_3-20160524-A.pdf.

Radu, L.-D. (2018). Environmental issues in internet of things: Challenges and solutions. *Acta Universitatis Danubius, Œconomica*, 14(1), 20–32 [Online]. Available at: http://journals.univ-danubius.ro/index.php/oeconomica/article/view/4250/4393.

Real-Time Innovations, Inc. (2013). Comprehensive Summary of QoS Policies (Cheat Sheet) [Online]. Available at: http://community.rti.com/rti-doc/500/ndds.5.0.0/doc/pdf/RTI_CoreLibrariesAndUtilities_QoS_Reference_Guide.pdf.

Richardson, L., Ruby, S., Hansson, D.H. (2007). *Restful Web Services*. 1st edition, O'Reilly Media [Online]. Available at: http://gen.lib.rus.ec/book/index.php?md5=569428a49f833a9ee3f50d0c010ab4c.0.

Taivalsaari, A. and Mikkonen, T. (2018). A taxonomy of iot client architectures. *IEEE Software*, 35(3), 83–88.

Vermesan, O., Friess, P., Guillemin, P., Gusmeroli, S., Sundmaeker, H., Bassi, A., Jubert, I.S., Mazura, M., Harrison, M., Eisenhauer, M., Doody, P. (2009). Internet of things strategic research roadmap. In *Internet of Things – Global Technological and Societal Trends*, Vermesan, O. and Friess, P. (eds). River Publishers, Aalborg.

Yang, J., Sandström, K., Nolte, T., Behnam, M. (2012). Data distribution service for industrial automation. *Proceedings of 2012 IEEE 17th International Conference on Emerging Technologies Factory Automation (ETFA 2012)*, 1–8.

Yuang, C. and Thomas, K. (2016). Performance evaluation of IoT protocols under a constrained wireless access network. *International Conference on Selected Topics in Mobile Wireless Networking (MoWNeT)*, 1–7.

Zhang, Q. and Fitzek, F.H. (2015). Mission critical IoT communication in 5G. In *Future Access Enablers of Ubiquitous and Intelligent Infrastructures*, Atanasovski, V. and Leon-Garcia, A. (eds). Springer, Cham.

6

Telerobotics

Wrya MONNET

Department of Computer Science and Engineering,
University of Kurdistan Hewlêr, Erbil, Iraq

6.1. Introduction

In the first chapters of this book, the Tactile Internet paradigm's primary enabling technologies, such as the low latency and the reliable communication network, were mentioned and explained. Both of these technologies are considered to be critical requirements to realize the haptic communication between the master and slave domains. In the Tactile Internet, the human is in the loop and informed about the slave side through tactile or kinesthetic feedbacks. These returns are used jointly or individually to get a real-time remote site experience, in order to control it efficiently and correctly.

In the network domain, malfunctioning, such as packet losses and increased transmission delays, may result in an unstable and less transparent system. Control capabilities of the system at the master and slave domains are necessary to compensate for the network domain inconveniences.

In a networked remote-controlled robotic system, with its sensors and actuators, many degrees of freedom (DoF) are required to enable an adequate level of dexterity. Machine learning algorithms and powerful embedded processors can provide good local control loops for better stability and transparency. Thus, to implement an overall teleoperated system, interdisciplinary techniques are involved, such as networking,

---

For a color version of all the figures in this chapter, see www.iste.co.uk/ali-yahiya/tactile.zip.

The Tactile Internet,
coordinated by Tara ALI-YAHIYA and Wrya MONNET. © ISTE Ltd 2021.

control systems, haptic sensors and machine learning. In this chapter, we tackle the effect of networking and the control system. We explore the possible architectures and control techniques in telerobotics and its evolution towards telepresence.

6.2. Teleoperation evolution to telepresence

In his article (Sheridan 1989), Sheridan debates whether it is the humans' or the robots' choice to perform tasks. After giving some salient considerations such as:

– unpredictable tasks are not doable by programmable machines,

– robots are preferable to perform work in hazardous environments,

– the lack of sufficient sensory information for controlling remote devices,

– the gradual and hard integration of artificial intelligence to improve remote-controlled devices,

he concludes that it is better to consider how robots and humans can collaborate instead of comparing them. Since then, teleoperation, telerobotics and telepresence have been successive evolutions of this concept of collaboration between human and robots.

Before more explanations, it is essential to mention the nuances between these three concepts (Sheridan 1995). In teleoperation, humans control remote sensors and actuators. A subclass of teleoperation is called telerobotics, in which humans supervise and control remote semiautomatic systems. Finally, Telepresence is the situation, in which special sensing and display technology enable the human to feel present at the remote location even though they are not there.

Historically, the remote-controlled techniques helped in working in hazardous environments and also in orthotics processes for manufacturing prosthetic organs, arms and legs. This gave rise to teleoperation technology, which is the extension of a person's sensing and manipulation capability to a remote location, in a master–slave bilateral manner. Initially, the remote location was "proximate" and in the operator's field of vision, known as "near-field" telerobotics. Later, this evolved to supervisory control (Tachi 1992), where the remote location was much farther and needed a telecommunication channel. The supervisory control was first handled in 1969 by Ferrell and Sheridan (1967), following the need for a remote-controlled manipulation system from Earth in space flight missions. The telerobotics concept has emerged with robotic techniques' progress, and merged with the supervisory control systems. Telerobotics with real-time presence sensations results in the concept of telepresence remote control, first mentioned by Minsky (1980). Another equivalent term for telepresence is "tele-existence", which was coined independently by Tachi in 1982 Tachi (1992) to describe the same type of systems. Throughout this book, we will use the term telepresence.

In the telerobotic system proposed by Ferrell and Sheridan (1967), as shown in Figure 6.1, two significant challenges faced by the authors were the communication time delay (three seconds round-trip for a lunar rover vehicle) and the remote environment that could not be directly observed from Earth. The suggested solution was a supervised remote autonomous computerized system capable of taking instantaneous local decisions while supervised remotely for necessary corrective measures. This system works as an open-loop where the operator is performing in a "move-and-wait" manner.

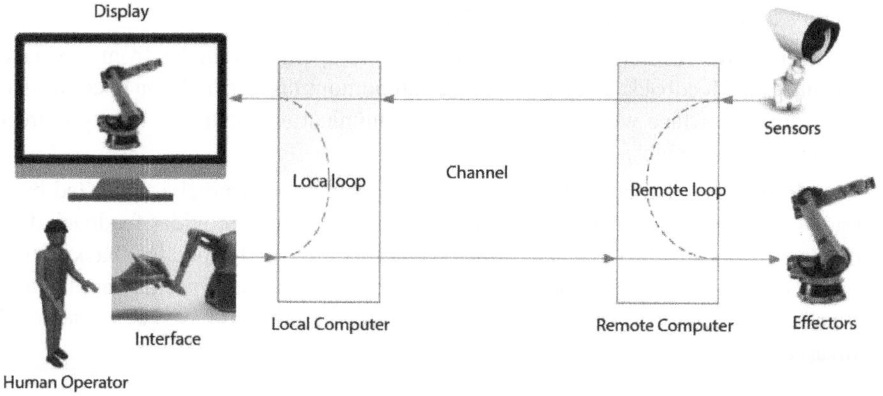

Figure 6.1. *Telerobotics as first devised by Sheridan (Ferrell and Sheridan 1967)*

Adding artificial intelligence to the supervisory control increases the flexibility and capability of the telepresence system. This helps to:

– overcome the limited dexterity of the operator;

– increase the remote system's capability of performing contingency operations;

– overcome unexpected channel malfunctioning, such as extended delays or reliability issues.

Telepresence can also be used in a virtual world (Tachi 1992) in addition to the real world. However, the virtual-world application has no network domain and slave domain, as shown in Figure 6.3, since a computer simulates the virtual world. The advantage of the virtual world is the possibility of interacting with the physical and non-physical world.

6.3. Telepresence applications

Human haptics perception, along with the computer haptic interface, is the way forward to immersion in virtual reality and telepresence. Another motivation for developing haptics is to provide a solution to computer interfaces for people who are blind and provide the feeling of immersion in virtual reality and the ability to feel virtual objects and surfaces. For example, the user's immersion in a remote environment happens by providing them with a local realistic environment that is similar. Many application domains are rising thanks to haptics, as shown in Figure 6.2. In this book, we will stick to telepresence, which, in turn, can be applied in different areas, such as:

– **space**: telepresence technology is useful in satellite retrieval and maintenance, deploying or assembling space platforms, scientific experiments and analysis of space materials in sealed space laboratories (Bejczy 1980);

– **underwater vehicles**: these are more remote-operated vehicles from a surface ship with video feedback. They can execute autonomous tasks in addition to controlled tasks from the surface with joysticks. The communication channel is through direct connection using cables;

– **telesurgery**: this consists of robotic motorized arms remotely controlled by a surgeon. The robotic arms interact with the environment and provides feedback data from the sensors attached to it. The surgeon receives force and haptic feedback so that they get close to the authentic environment on the remote side. Since this application is a critical one, only a few long-distance telesurgery operations have been conducted (Anvari *et al.* 2005);

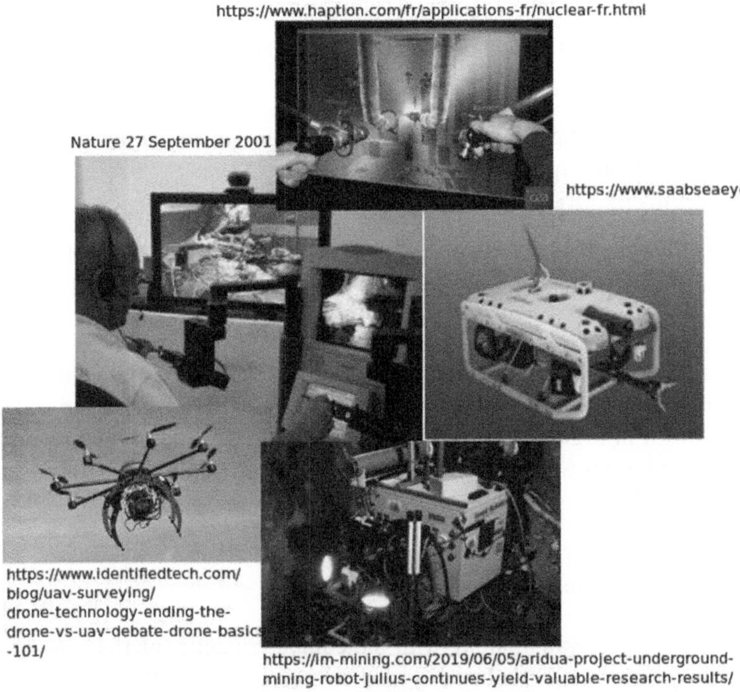

Figure 6.2. *Different telepresence applications*

– **nuclear field**: telerobotics is used to avoid exposing humans to a dangerous radioactive area in the nuclear field and prevents exposure to hazardous radioactive materials. The haptic feedback of force is necessary because of the dangerousness of the elements involved and the associated risk when handling them;

– **terrestrial mining**: the hazardous environment and the heavy equipment used in mining require the use of remotely operated robots. Equipment, such as excavating machines and exploration robots, can be operated from the surface to do drilling and explore the unreachable places in order to take samples;

– **social life and entertainment**: with telepresence, the application of social communication with the sensation of being present can be realized. The touch and haptic feedback are the essential requirements of such applications. An example of such a system is implemented in Eid *et al.* (2008), where a haptic exchange happens during videoconferencing;

– **unmanned aerial vehicles (UAVs)**: UAVs are teleoperated robots used in various situations, including disaster area inspection and movie content creation (Keita Higuchi 2013).

In the following section, the different components of the teleoperation system are presented. The use of the term teleoperation is for generality.

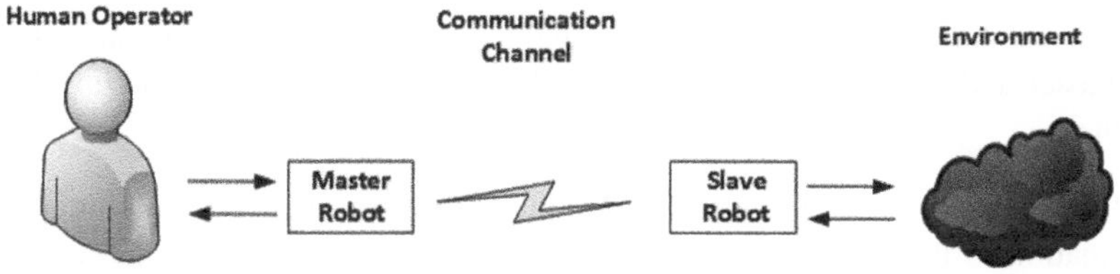

Figure 6.3. *Components of a telerobotic*

6.4. Teleoperation system components

Teleoperation or "operating at a distance" is defined by Sheridan as a system that "extends a person's sensing and/or manipulating capability to a location remote to that person" (Sheridan 1989). A teleoperator is a system, on the remote side, consisting of sensors and mechanisms to sense the environment and to realize the commands sent by a remote human operator.

Telepresence or tele-existence is a teleoperated system equipped with an advanced sensing system to convey the haptic information to the human operator through a reliable and low latency communication channel. The human operator feels almost as though they were physically present at the remote side. To do so, the operator needs a haptic interface to transform the haptic signals sent by the teleoperator.

Figure 6.3 shows the different components of a teleoperated system. It consists of the human operator, the haptic interface on the master side, the remote robot on the slave side, the environment and the communication network:

– The human operator who initiates the position and force commands to the remote environment.

– The haptic interface consists of the sensor and actuators that, on the one hand, transform the master operator's movements into position and force signals, and, on the other hand, reproduces the remote robot's kinesthetic and tactile information to the operator. A survey of up-to-date available glove-like haptic devices is given in (Dipietro *et al.* 2008; Wang *et al.* 2019). Other commercial haptic interfaces are also available, some of which are shown in Figure 7.4. While the wearable haptic interface is not highly developed commercially, some researchers are working on their development (Lei *et al.* 2019). In section 7.3.1, we will give detailed information about the working principles of tactile sensors and actuators.

– The remote robot slave consists of a robot that acts on the remote environment with some sensors to feedback the position, force and tactile information to the operator.

– The environment is the medium on which the actions are exerted. This can be stiff or flexible objects, significant physical obstacles or human, depending on the application in which it is applied.

– The communication network, the medium in which all the information between the master and slave is exchanged. It can be wired, wireless or made up of optical channels.

Teleoperation can be unilateral where the master transmits the position and force information to the slave side without feedback from the slave to master. This is a classical open-loop control between the human operator who sends the commands and the slave system. The slave has its local closed-loop control for better performance. In a bilateral teleoperation system, the slave feeds information such as haptic and motion and/or force back to the master. Both sides of the teleoperation system have local controls to manage the position, velocity and/or forces. A computerized system on both sides can do this.

Telerobotic system control architecture can be classified into three main categories: direct control, shared control or supervisory control (Sheridan 1989, 1992).

In a direct control telerobotics architecture (also called manual control), the human operator is in full control of the remote task, i.e. the system has no intelligence or autonomy. In this case, the remote robot tools are not automated but are teleoperated systems under the operator's direct control.

In a shared control architecture, a computerized robotic system is used on both the operator and teleoperator sides to augment the human operator's capability with their sensory feedback systems. An example is the telerobotic systems used in telesurgery, where the surgeon is in full control of the task. However, the computer control system can add more precision to the realized task.

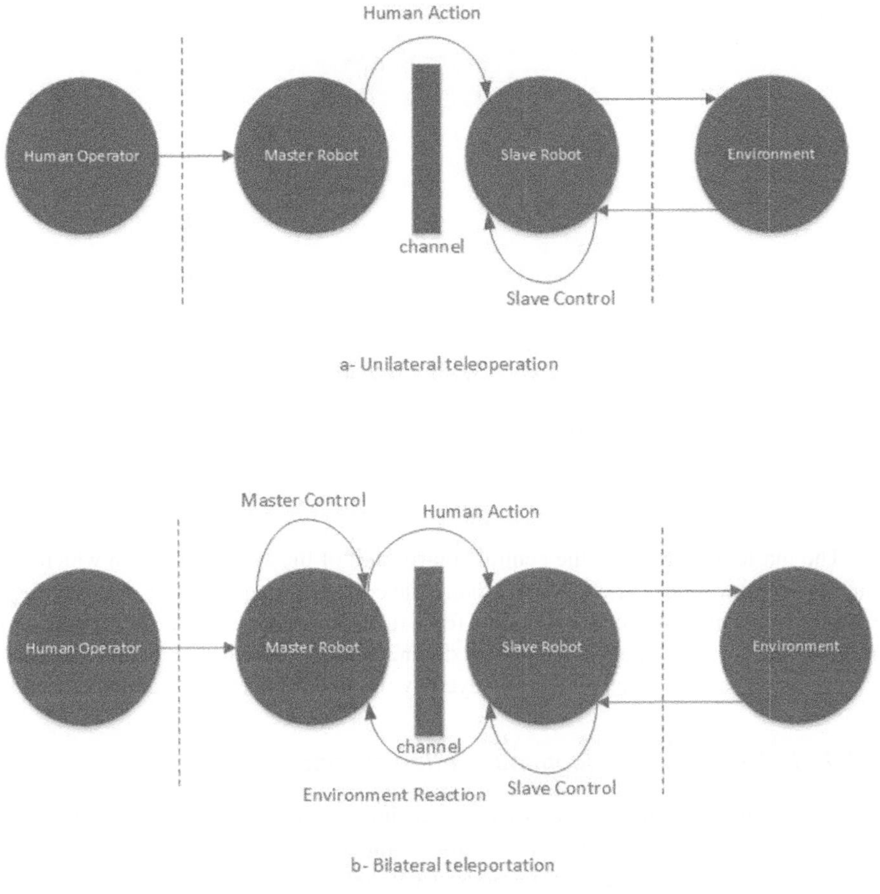

Figure 6.4. *Unilateral and bilateral teleoperation*

In a supervisory control architecture, humans supervise a semi-autonomous system in these telerobotic systems, such as the auto-pilot of an aircraft or a chemical or power plant. This is analogous to a human supervisor directing and monitoring the activities of a human subordinate.

The teleoperation system comprises of different parts: the master domain and the communication link slave domains, as shown in Figure 6.5. These components are linked together in many configurations (architectures), depending on the exchanged signals (forward and feedback) between the master and slave domains. Details on these architectures will be given in section 6.5.

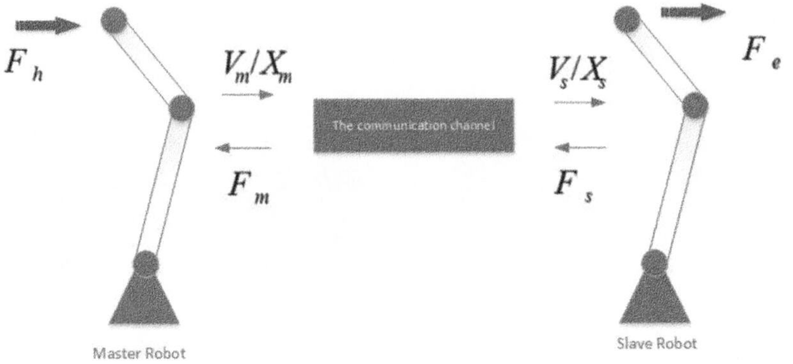

Figure 6.5. *Teleoperation system with the three domains with robotic arm and haptic interface, where X_m and X_s stand for the master and slave position variables, respectively*

6.4.1. *Master domains*

The master controller is the central control unit of the system, through which the human operator sends the commands to the slave via a human interface. The interface transforms the human operator's exerted force to position and velocity information and receives the force feedback. The master domain's tactile interface contains actuators to receive and convert the haptics conveyed by the slave side to the operator.

6.4.2. *Network domain (communication channel)*

The communication layer serves as the medium of information exchange between the master and slave components. This communication support ranges from a dedicated link, such as cables in underwater vehicles, to a network connection, such as the Internet or wireless communication systems. The teleoperated system's performance and stability mainly depend on the communication link's reliability and transmission delay.

6.4.3. *Slave domain*

The slave domain consists of actuators, sensors and a local feedback system to improve the mechanical system's position and velocity control. The actions of the slave component are exerted on the environment on the remote side. Different kinds of sensors, including tactile and haptic sensors, are embedded in the slave interface to improve control and to feedback tactile and force signals to the master domain.

6.5. Architecture of bilateral teleoperation control system

In section 6.4, the teleoperation components master domain, network domain (communication channel) and slave domain were briefly explained. This section presents a two-port model of the master and slave domains and the communication channel. Both domains consist of control parts for better stability and transparency of the telerobotic system. Hence, the mathematical model for these different control parts is essential for design purposes.

In Hannaford (1989), the author suggested a framework for this design by modeling the whole system using a two-port linear system, which takes into account the bilateral exchange of energy between the human operator and the input port, and between the manipulator and the environment. Figure 6.6 shows this model for a one degree of freedom master–slave teleoperation system. The dynamics of this system is characterized by a set of impedances. The variables are the force and velocity, equivalent to the generalized quantities "effort" and "flow", respectively. These variables are found at the input and output ports since the energy is exchanged from both ends. The general terms can also represent "displacement" and "vibration," or "torque" and "angular velocity". A physical representation of this two-port model is shown in Figure 6.6. M_m, M_s, f_m and f_s are the master and slave inertia and control signals, respectively. The master Z_m impedance is the haptic display impedance. It generates mechanical impedance, where "impedance" is defined as a dynamic relationship between velocity and force, with f_h the force exerted by the operator's hand on the master and f_e the force exerted by the environment on the slave. A haptic display's capacity to render a range of dynamic range of impedances while keeping the system passive is called the Z-Width. A display with a larger Z-Width will usually render a better feeling of the slave environment. Hence, Z-Width may be viewed as a measure of quality for the haptic display (Colgate and Brown 1994). F_h^* and F_e^* are the operator's and the environment's exogenous input forces and are independent of the teleoperation system behavior. Z_h and Z_e denote the dynamic characteristics of the human operator's hand and the remote environment, respectively. Finally, the transmitted impedance Z_t is the user's perception of the environment impedance Z_e and Z_r is the operators' impedance viewed from the environment end.

By choosing the output force f_e and the input velocity v_h as dependent variables, the two-port network can be represented in the s-domain by the following equation (Hannaford 1989):

$$\begin{bmatrix} F_h(s) \\ -V_e(s) \end{bmatrix} = \begin{bmatrix} h_{11} & h_{12} \\ h_{21} & h_{22} \end{bmatrix} \begin{bmatrix} V_h(s) \\ F_e(s) \end{bmatrix} \qquad [6.1]$$

where

$$h_{11} = \frac{F_h(s)}{V_h(s)}|_{F_e(s)=0} = Z_t \,, \qquad h_{22} = -\frac{V_e(s)}{F_e(s)}|_{V_h(s)=0} = \frac{1}{Z_r}$$

where Z_t is the impedance of the two-port network at the master end and Z_r is the impedance of the two-port from the slave end. Hence, h_{22} signifies admittance. Also,

$$h_{12} = \frac{F_h(s)}{F_e(s)}|_{V_h(s)=0}, \qquad h_{21} = -\frac{V_e(s)}{V_h(s)}|_{F_e(s)=0}$$

The parameter h_{12} is a measure of force tracking under hard contact since $V_h(s)=0$ and h_{21} is a measure of velocity tracking in free space since $F_e(s) = 0$.

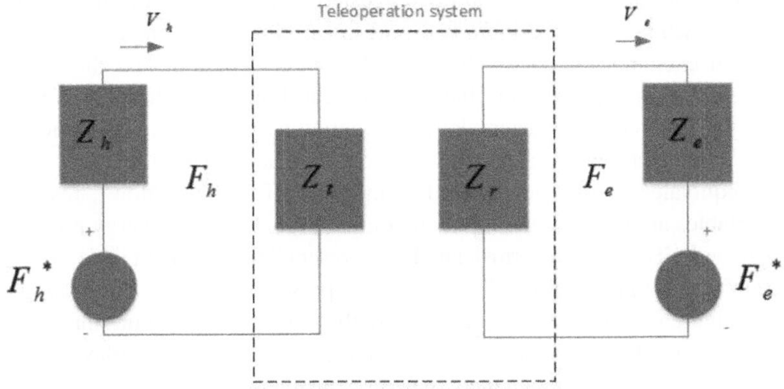

Figure 6.6. *Two-port model of the teleoperation system*

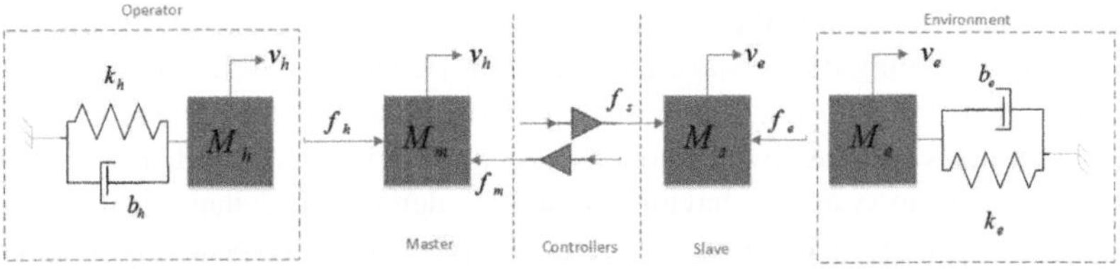

Figure 6.7. *Mechanical model of the two-port teleoperation system*

The parameters h_{ij} are affected by the mechanical dynamics of the master and slave as well as the control architecture. A realistic physical representation of the master–slave system is shown in Figure 6.7. Based on this, a general control architecture can be suggested for analysis and design purposes of a bilateral teleoperation system (Lawrence 1993), as shown in Figure 6.8. The control architecture includes the dynamics of the master and slave mechanisms Z_m and Z_s, along with their local closed-loop control subsystems using proportional and derivative PD controls C_m and C_s, respectively.

A more specific control architecture can be derived from the general one in Figure 6.8. A position–velocity architecture, also known as direct force reflection (DFR), is obtained by removing some controls and communication subsystems, while keeping others. Figure 6.9 shows this control architecture, where C_4 and C_3 are removed, $C_1 = C_s$ and $C_2 = 1$, $Z_m = M_m s$ and $Z_s = M_s s$.

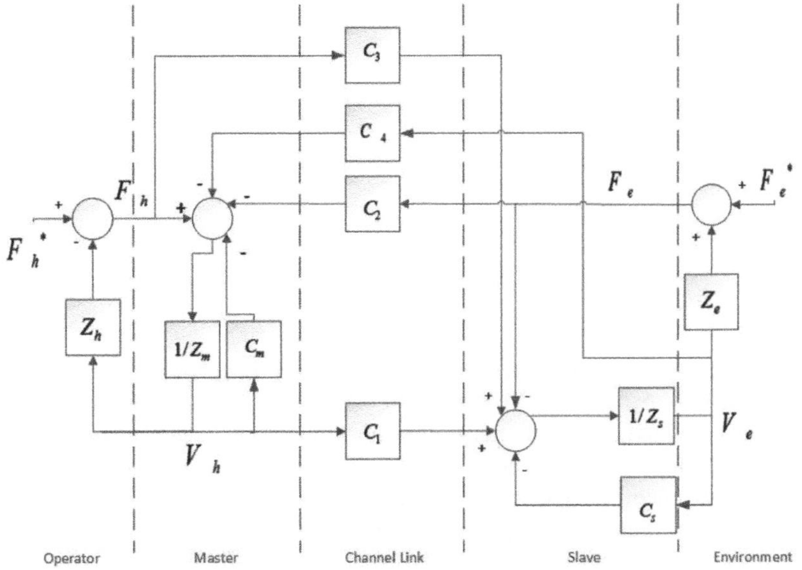

Figure 6.8. *General four-channel bilateral teleoperator system architecture (Lawrence 1993)*

Referring to Figure 6.9 and the parameter in equation [6.1], and knowing that $F_e(s) = Z_e V_e$ from Figure 6.6 with $F_e^* = 0$ for passivity reasons, the transfer function from $F_h(s)$ to $V_h(s)$ is given by

$$\frac{V_h}{F_h} = \frac{1 + h_{22} Z_e}{h_{11}(1 + h_{22} Z_e) - h_{12} h_{21} Z_e} \qquad [6.2]$$

To analyze the system's stability, seen from the master end, the zeros and poles of the transfer function's characteristic equation are to be considered. By assuming the environment to be a linear spring, $Z_e = k_e$, where k_e is its stiffness, and the characteristic equation is given by

$$h_{11} s + k_e (h_{11} h_{22} - h_{12} h_{21}) = 0 \qquad [6.3]$$

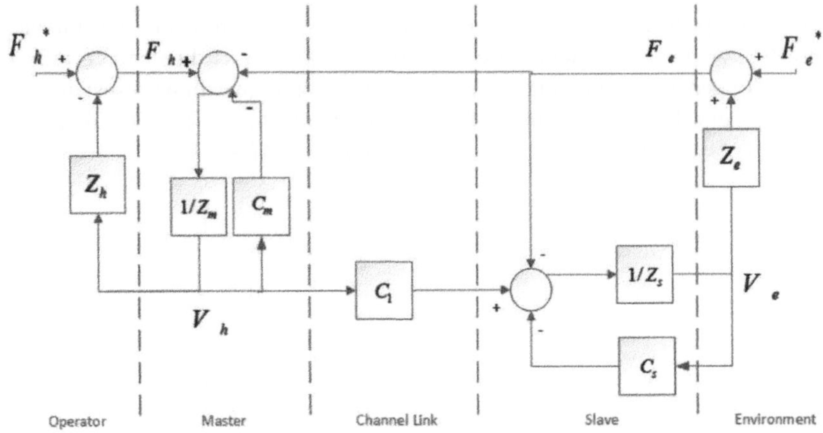

Figure 6.9. *Direct force reflection architecture*

6.5.1. *Classification of the control systems architectures*

In the previous section, a general architecture of the teleoperated system is presented as a four-channel control system, where each channel had its own control law C_i's (Lawrence 1993). Control architectures for bilateral teleoperation are often classified by the number of signals transmitted over the communication channel. The general four-channel architecture is also called the position–force–position–force (PF–PF) architecture, since both flows are communicated in both directions. Other control classes of the bilaterally teleoperated system are obtained by selecting the different subsystems $(C_1 - C_6)$. For example, a position–position architecture is a two-channel architecture and P–PF is a three-channel architecture.

Figure 6.9 shows the diagrams of the P–F (also called the direct force reflection (DFR)) architecture. Similarly, other classes, such as the P–P (also called the position error based (PEB)) architecture also exist. The two-channel P–P architecture performance can still be improved by adding a third channel, for example, P–PF or PF–P. Another three-channel architecture is the DFR, with the forward force and feedback force, i.e. PF–F. More works on PF–F architectures have been published in comparison to other architectures, due to the various disadvantages of the other combinations, such as delay-introduced forces and lack of position tracking (Heck 2015). A common problem in DFR is that directly feeding back contact forces results in a violent master reaction when the slave contacts a stiff environment. Hence, to stabilize the system, the reflected forces' magnitude must be attenuated (Daniel and McAree 1998). In the following section, the PF–F architecture is analyzed in its discrete-time model.

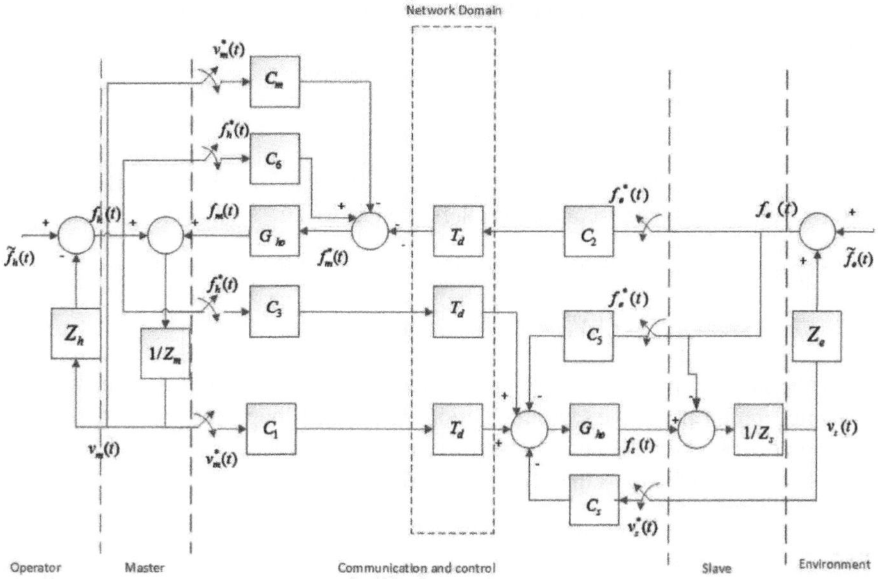

Figure 6.10. *Discrete position force–force architecture*

6.5.2. *Discrete architecture with transmission delay*

The bilateral control architecture explained above is mainly used in a geographically distant place with a communication channel, typically based on Internet or 5G links. Therefore, on the one hand, the force and velocity signals in Figure 6.9 should be sampled and coded before being passed to the network. On the other hand, there will be a delay observed on the other end of the communication channel resulting from the core network and the protocols, in addition to the propagation delay. The sampling frequency of the force and velocity is a critical parameter to choose: it is essential in capturing authentic haptic effects as well as playing a role in the stability of the system (Tavakoli *et al.* 2008). Consequently, the sampling frequency will affect the bandwidth of the channel used for communication.

An analysis of the system operation is given below, taking into account a negligible transmission delay. In section 6.6.2, the time delay of transmission is added, to find its effect. The local hand controller and the remote teleoperated slave parts in Figure 6.9, being the mechanical parts, will remain, while the force and velocity signals, passed through the communication channel, are sampled on one end and reconstructed on the other. This is shown in Figure 6.10, where f_m^* and v_h^* are the discrete counterparts of f_m and v_h, respectively. $G_{h0}(s)$ is the transfer function of a zero-order hold (ZOH) block to reconstruct a continuous signal from the discrete ones. It is given, in the Laplace domain, as follows

$$G_{h0} = \frac{1 - e^{sT}}{s}$$

where T is the sampling time, considering the discrete-time control signals, and, taking C_m and C_s to be discrete controllers, we can write:

$$\begin{aligned} F_m^* &= -C_2 F_e^{*T_d} - C_m V_m^* + C_6 F_h^* \\ F_s^* &= C_1 V_m^{*T_d} - C_s V_s^* - C_5 F_e^* + C_3 F_h^{*T_d} \end{aligned} \quad [6.4]$$

where the superscript T_d stands for the delayed version of the sampled variable, and it is an integer multiple of the sampling period. The closed-loop dynamics of the system in Figure 6.10 in the z domain can be rewritten by taking the force as an independent variable and the velocity as the dependent variable.

$$\begin{aligned} V_m(z) &= \mathcal{Z}[Z_m^{-1} F_h(t)] + \mathcal{Z}[Z_m^{-1} G_{h0}](-C_2 F_e(z) z^{-T_d} - C_m(z) V_m(z) \\ &\quad + C_6 F_h(z)) \\ V_s(z) &= -\mathcal{Z}[Z_s^{-1} F_e(t)] + \mathcal{Z}[Z_s^{-1} G_{h0}](C_1 V_m(z) z^{-T_d} + C_3 F_h(z) z^{-T_d} \\ &\quad - C_s(z) V_s(z) - C_5 F_e(z)) \end{aligned} \quad [6.5]$$

where

$$\mathcal{Z}[Z_m^{-1} G_{h0}] = \frac{T}{M_m(z-1)} \text{ and } \mathcal{Z}[Z_s^{-1} G_{h0}] = \frac{T}{M_s(z-1)} \quad [6.6]$$

where M_m and M_s are master and slave inertias, respectively. To find the Z-transform of the terms $Z_s^{-1} F_e(t)$ and $Z_m^{-1} F_h(t)$, we use the approximation method given in (Jury 1964), where, to find the Z-transform of an arbitrary signal, it is passed through an integrator operator $G(s) = 1/s^n$, after a sampling and a reconstruction using an interpolator

$$F_{h_a}(z) = F_h(z) \mathcal{Z}[QG]$$

where Q is the transfer function of the interpolator, G is the integrator operator and the subscript $_a$ stands for the approximation. Using a quadratic interpolator and a one-stage integrator, the book's tables give the following approximation:

$$F_{h_a}(z) = F_h(z) \frac{T(1 + 4z^{-1} + z^{-2})}{3M_m(1 - z^{-2})} \quad [6.7]$$

By substituting equations [6.6] and [6.7] into [6.5], we obtain the hybrid model of a discrete bilateral teleoperation system

$$V_m(z) = \frac{T(1+4z^{-1}+z^{-2})}{3M_m(1-z^{-2})}F_h(z) + \frac{T}{M_m(z-1)}$$
$$(-C_2F_e(z)z^{-T_d} - C_m(z)V_m(z) + C_6F_h(z))$$
$$V_s(z) = -\frac{T(1+4z^{-1}+z^{-2})}{3M_s(1-z^{-2})}F_e(z) + \frac{T}{M_s(z-1)}$$
$$(C_1V_m(z) - C_s(z)V_s(z) + C_3F_h(z)z^{-T_d} - C_5F_e(z)) \qquad [6.8]$$

Rearranging these two equations and by defining new parameters as a function of the sampling period T, inertia M_i and polynomials in z, we get:

$$\zeta = \frac{T(z^2+4z+1)}{3M_m(z^2-1)}\gamma = \frac{T}{3M_m(z^2-1)}$$

The input force F_h and slave velocity V_s can be rewritten as functions of feedback force F_e and input velocity V_m, as shown in the following equation:

$$F_h(z) = \frac{(1+C_m(z)\gamma)}{\zeta + C_6\gamma}V_m(z) + \frac{C_2\gamma z^{-T_d}}{\zeta + C_6\gamma}F_e(z)$$
$$V_s(z) = -\left(\frac{\gamma C_1 z^{-T_d}}{1+\gamma C_s(z)} - \frac{\gamma C_3(1+C_m(z)\gamma)z^{-T_d}}{(1-\gamma C_s(z))(\zeta + C_6\gamma)}\right)V_m(z)$$
$$\left(\frac{\gamma^2 C_3 C_2}{(1-\gamma C_s(z))(\zeta + C_6\gamma)} - \frac{C_5\gamma}{1+\gamma C_s(z)}\right)F_e(z) \qquad [6.9]$$

This is, in fact, the hybrid model of the discrete controlled teleoperated system.

$$\begin{bmatrix} F_h(z) \\ V_s(z) \end{bmatrix} =$$
$$\begin{bmatrix} \frac{(1+C_m(z)\gamma)}{\zeta + c_6\gamma} & \frac{C_2\gamma z^{-T_d}}{\zeta + C_6\gamma} \\ -\left(\frac{\gamma C_1 z^{-T_d}}{1+\gamma C_s(z)} - \frac{\gamma C_3(1+C_m(z)\gamma)z^{-T_d}}{(1-\gamma C_s(z))(\zeta+C_6\gamma)}\right) & \left(\frac{\gamma^2 C_3 C_2}{(1-\gamma C_s(z))(\zeta+C_6\gamma)} - \frac{C_5\gamma}{1+\gamma C_s(z)}\right) \end{bmatrix} \begin{bmatrix} V_m(z) \\ F_e(z) \end{bmatrix}$$
$$[6.10]$$

Two instability sources are present in this model: the delay in the transmission line T_d and the discretization at a period of T. The delay in transmission will change the overall system from passive to active (see section 6.6.1 and section 6.6.2). This is due to the power generation in the transmission line and the ZOH. To overcome this instability, in Anderson and Spong (1989), the author presented the scattering matrix approach to ensure the passivity of a telerobotic system, with arbitrary transmission delay, for a stable system operation. Another stabilization method, by using the wave variable, was introduced by Niemeyer and Slotine (1991) to a bilateral teleoperation system with transmission delay. The discretization in time is another source of instability for the bilateral system. In Tavakoli *et al.* (2008), the effect of sampling period T on stability is found by replacing the $h_{ij}(z)$ parameters in the characteristic equation of the master subsystem given in [6.3]. Considering no transmission delay, $T_d = 0$, we find the effect of T. By finding the characteristic equation's roots, with the poles on the right-hand side of the s-plane for a stable system, a relationship between T, environment stiffness and control parameters, such as damping factors, is found. It is such that, when the sampling period is increased for stable interaction with the environment, the maximum stiffness with which a slave robot can work is reduced.

6.6. Performance and transparency of telepresence systems

The transparency of a telepresence system is measured by how much the master domain can reproduce the slave environment accurately. A non-experienced operator, for example, cannot manipulate a non-transparent telerobotic system correctly because of inaccurate returns from the remote side. Therefore, a good haptic interface and a passive network with reduced delay and no jitter are ideally needed for a good performance in haptic systems. Many other factors affect the telerobotic systems' performances: the operator's experience, skills and reaction speed, the update rate of the haptic information, the environment sensors (their dynamic ranges) and the haptic interface speed and accuracy. In Skubic *et al.* (1995), some metrics, based on physical value, such as position error, distance and time, are suggested and tested to measure accuracy and speed. The endurance and smoothness of the system are also considered as performance metrics.

The transparency of the telerobotic system is defined by Lawrence (1993) as the extent to which the impedance in the domain of the remote slave is transmitted to or felt by the operator on the master domain. It it has to be equal to the impedance of the remote environment, i.e. when the impedance perceived by the operator Z_h is equal to the environment impedance Z_e in Figure 6.6. Ideally, the transfer function of equation [6.2] should give $1/Z_e$ for perfect transmission of the admittance of the slave side. An objective measure of haptic displays is "Z-Width", which represents the dynamic range of impedances that can be passively rendered. Haptic displays with larger Z-Width generally cause a more realistic feeling of remote slave and virtual environments.

6.6.1. *Passivity and stability*

A passive system does not produce energy internally but dissipates it. A stable system is when a given bounded input is introduced to it, and then it has a bounded output. Hence, it can be intuitively stated that a passive system is stable. Transparency in the teleoperated system, as defined before, is the extent to which the slave environment is mirrored at the master haptic interface. A more detailed explanation and the conditions for a system to have both properties are explained below.

Passivity is the property of many physical systems; it reveals their stability from the input/output energy perspective. The input power to a system is equal to the system's input and output vectors' inner product. For example, the input vector of the two-port system in Figure 6.6 is the force vector F_h, the output vector is the velocity vector V_e and the input power is the product of force and velocity $P_{in} = V_h^T F_h$. This, in turn, will be equal to the variation concerning the time of stored energy in the system E_{stored} and the dissipated power P_{dissp},

$$P_{in} = V_h^T F_h = \frac{dE_{sored}}{dt} + P_{dissp} \qquad [6.11]$$

By integrating over a time t, the initial stored energy term $E_{stored}(0)$ in the system will appear. Its value is negative, since the system supplies this energy. To satisfy equation [6.11], the input energy should be greater than or equal to the stored energy in the system:

$$\int_0^t P_{in} dt = \int_0^t V_h^T F_h dt = E_{sored} - E_{stored}(0) + \int_0^t P_{dissp} dt \qquad [6.12]$$

In a passivity-based system, a system's output energy will not exceed the sum of the input energy and the system's initial stored energy.

If the power dissipation is zero for all time, the term for this type of system is lossless. In contrast, if the power dissipation is positive, as long as the stored energy has not reached its lower bound, the system is strictly passive. Passivity is a sufficient condition for stability; hence, a passive teleoperation system is guaranteed to be stable. However, a stable system is not necessarily passive.

Passivity is related to stability since in a passive system we cannot store more energy than supplied by a "source". From an energetic viewpoint, a system is passive if it absorbs more energy than it produces: it can store and dissipate energy but will not amplify it.

A trade-off between stability and transparency is an inherent problem in teleoperation systems. This is due to the bilateral teleoperation property, where stability and passivity are not always guaranteed because of the force feedback (see section 6.6.2). Moreover, instability, caused by the transmission delay, will also affect the transparency. In the following section, two approaches to mitigating the transmission delay are given.

6.6.2. *Time delay issues*

In telepresence tactile systems, the force is feedback from a remote slave device to the master operator. Delayed force feedback imposes disturbance on the operator's hand, who cannot ignore and react to it. This, in turn, will cause instability in the process.

To analyze this effect, in a two-port system and linear time invariant (LTI), we use the scattering theory, initially developed to analyze transmission line systems. It consists of an operator S in the s-domains that maps the input (effort + flow) to the output (effort - flow), as given below (Anderson and Spong 1989):

$$F(s) - V(s) = S(s)[F(s) + V(s)] \quad [6.13]$$

where $F(s)$ is the effort applied across the system and $V(s)$ is the flow injected into the system in the Laplace domain. The equation states that the scattering operator maps effort plus flow (incident wave) into effort minus flow (reflected wave). The flow enters the system's ports, and effort is measured across the system's ports. According to the passivity principle, the reflected wave cannot be greater than the incident one; therefore, the S operator should be less than or equal to 1, where the force and velocity in the s-domain, $F(s)$ and $V(s)$, respectively, are square-integrable (Anderson and Spong 1989). For the two-port system in equation [6.1], the scattering matrix $S(s)$ can be related to the hybrid matrix $H(s)$, using the notations in [6.6] as follows:

$$\begin{bmatrix} F_h(s) - V_h(s) \\ F_e(s) + V_e(s) \end{bmatrix} = \begin{bmatrix} 1 & 0 \\ 0 & -1 \end{bmatrix} \left(\begin{bmatrix} F_h(s) \\ -V_e(s) \end{bmatrix} - \begin{bmatrix} V_h(s) \\ F_e(s) \end{bmatrix} \right) \quad [6.14]$$

$$= \begin{bmatrix} 1 & 0 \\ 0 & -1 \end{bmatrix} (H(s) - I) \begin{bmatrix} V_h(s) \\ F_e(s) \end{bmatrix} \quad [6.15]$$

Likewise, from the other side of the two-port network:

$$\begin{bmatrix} F_h(s) + V_h(s) \\ F_e(s) - V_e(s) \end{bmatrix} = \begin{bmatrix} F_h(s) \\ -V_e(s) \end{bmatrix} + \begin{bmatrix} V_h(s) \\ F_e(s) \end{bmatrix} = (H(s) + I) \begin{bmatrix} V_h(s) \\ F_e(s) \end{bmatrix} \quad [6.16]$$

From the two equations above, we find that matrix S can be written as a function of the hybrid matrix as:

$$S(s) = \begin{bmatrix} 1 & 0 \\ 0 & -1 \end{bmatrix} (H(s) - I)(H(s) + I)^{-1} \quad [6.17]$$

The authors in Anderson and Spong (1989) found that the necessary and sufficient condition of passivity in the teleoperation system, as in equation [6.1], is that the scattering matrix's norm must not exceed 1. In other words, the scattered wave cannot have an energetic content greater than the incident wave. By letting the hybrid matrix $H(s)$ introduce only time delay operators, it is found that the scattering matrix is not bounded; hence, the system is not passive and unstable.

Adding a corrective module to the transmission line such that the norm of the scattering matrix (or the highest eigenvalues) becomes $\|S\| \leq 1$, the transmission line system becomes passive and stable. It is shown that the following scattering matrix will make the system passive:

$$S(s) = \begin{bmatrix} 0 & e^{-sT} \\ e^{-sT} & 0 \end{bmatrix} \qquad [6.18]$$

By considering Figure 6.6 and applying the above scattering, we can rewrite:

$$\begin{bmatrix} F_h(s) - V_h(s) \\ F_e(s) + V_e(s) \end{bmatrix} = \begin{bmatrix} 0 & e^{-sT} \\ e^{-sT} & 0 \end{bmatrix} \begin{bmatrix} F_h(s) + V_h(s) \\ F_e(s) - V_e(s) \end{bmatrix} \qquad [6.19]$$

By transforming equation [6.19] into the time domain:

$$\begin{bmatrix} F_h(t) - V_h(t) \\ F_e(t) + V_e(t) \end{bmatrix} = \begin{bmatrix} F_e(t-T) - V_e(t-T) \\ F_h(t-T) + V_h(t-T) \end{bmatrix} \qquad [6.20]$$

The passive control law for the communication system, without the scaling factor between the velocity and force, is then

$$F_h(t) = F_e(t-T) - V_e(t-T) + V_h(t) \qquad [6.21]$$

$$V_e(t) = V_h(t-T) - F_e(t) + F_h(t-T) \qquad [6.22]$$

Equations [6.21] and [6.22] show that the scattering matrix maps the relationship between the velocity and force signals and their delayed signals to the system. Finally, since the force and velocity signals may differ by orders of magnitude because of different units, scaling is necessary for implementation reasons.

A conceptually similar formulation to the scattering formulation is the wave-variable formulation. It is based on relating the velocity and force by a parameter b, defined as the characteristic impedance or wave impedance, as shown in Figure 6.11. New variables, u_m and v_s, are generated, to be transmitted across the network. These variables are given by

$$u_m(t) = \frac{1}{\sqrt{2b}}(F_h(t) + bV_h(t)) \qquad [6.23]$$

$$u_s(t) = \frac{1}{\sqrt{2b}}(F_s(t) - bV_e(t)) \qquad [6.24]$$

$$v_s(t) = \frac{1}{\sqrt{2b}}(F_s(t) + bV_e(t)) \quad [6.25]$$

$$v_m(t) = \frac{1}{\sqrt{2b}}(F_h(t) - bV_h(t)) \quad [6.26]$$

The transformation is bijective; therefore, by using the wave variables u_m, v_m, u_s and v_s and given the input and output forces, the velocity can be found as:

$$F_h(t) = -bV_h(t) + \sqrt{2b}v_s(t) \quad [6.27]$$

$$F_s(t) = bF_s(t) + \sqrt{2b}v_s(t) \quad [6.28]$$

$$V_h(t) = \frac{1}{b}(F_h - \sqrt{2b}v_m) \quad [6.29]$$

$$V_e(t) = -\frac{1}{b}(F_s - \sqrt{2b}v_s) \quad [6.30]$$

It can be shown that lossless passive communication can be achieved by directly transmitting the wave variables u and v instead of F and V (Niemeyer and Slotine 1991). Moreover, the passivity property is entirely independent of the actual time delay. This is shown by measuring the power input to the transmission line in Figure 6.11, knowing that:

$$v_m(t) = u_s(t-T) \quad v_s(t) = u_m(t-T) \quad [6.31]$$

$$P_{in} = \frac{1}{2}u_m(t)^2 - \frac{1}{2}v_m(t)^2 + \frac{1}{2}u_s(t)^2 - \frac{1}{2}v_s(t)^2 \quad [6.32]$$

$$P_{in} = \frac{1}{2}u_m(t)^2 - \frac{1}{2}u_s(t-T)^2 + \frac{1}{2}u_s(t)^2 - \frac{1}{2}u_m(t-T)^2 \quad [6.33]$$

By integrating the power over the transmission delay time,

$$E_{stored}(t) = \int_{t-T}^{t} \frac{1}{2}u_m(\tau)^2 + \frac{1}{2}u_s(\tau)^2 d\tau \quad [6.34]$$

This means that the input power during the time delay is saved as energy, and condition [6.12] is satisfied. Moreover, the dissipated power is 0 and the system is lossless. Therefore, the communication channel has become a temporary energy storage element by using the wave-variable approach. Because of this, it can maintain passivity and be stable for an arbitrary time delay.

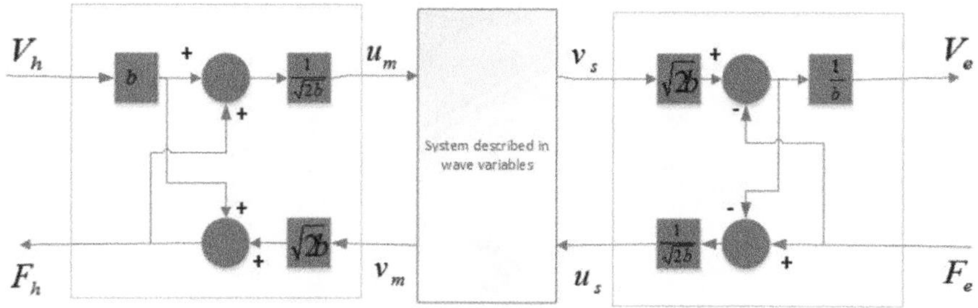

Figure 6.11. *Wave variable architecture to absorb time delay in the transmission medium*

The wave variables in [6.23] provide the same control law as in the scattering matrix case, by using, along with equations [6.31]:

$$\frac{1}{\sqrt{2b}}(F_h(t) - bV_h(t)) = \frac{1}{\sqrt{2b}}(F_s(t-T) - bV_e(t-T)) \qquad [6.35]$$

$$\frac{1}{\sqrt{2b}}(F_s(t) + bV_e(t)) = \frac{1}{\sqrt{2b}}(F_h(t-T) + bV_h(t-T)) \qquad [6.36]$$

which is:

$$F_h(t) = F_s(t-T) + b(V_h(t) - V_e(t-T)) \qquad [6.37]$$

$$V_e(t) = \frac{1}{b}(F_h(t-T) - F_s(t)) + V_h(t-T) \qquad [6.38]$$

which are equivalent to the control laws in [6.21] and [6.22], with the scaling factors b and $1/b$ added here.

Additional improvement of the wave transmission can be obtained by adding matched termination on both the master and slave sides to reduce the wave reflections in the force–velocity control system, as shown in Figure 6.12.

The author (Aziminejad *et al.* 2008) proposed wave theory stability for four-channel and three-channel bilateral telerobotic system for time-delay compensation (Figure 6.13). It is shown that ideal transparency is achieved in the presence of time delay.

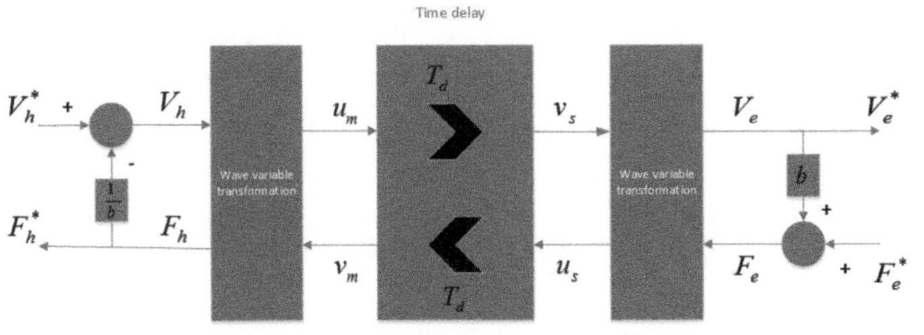

Figure 6.12. *Matched termination to improve wave-variable method*

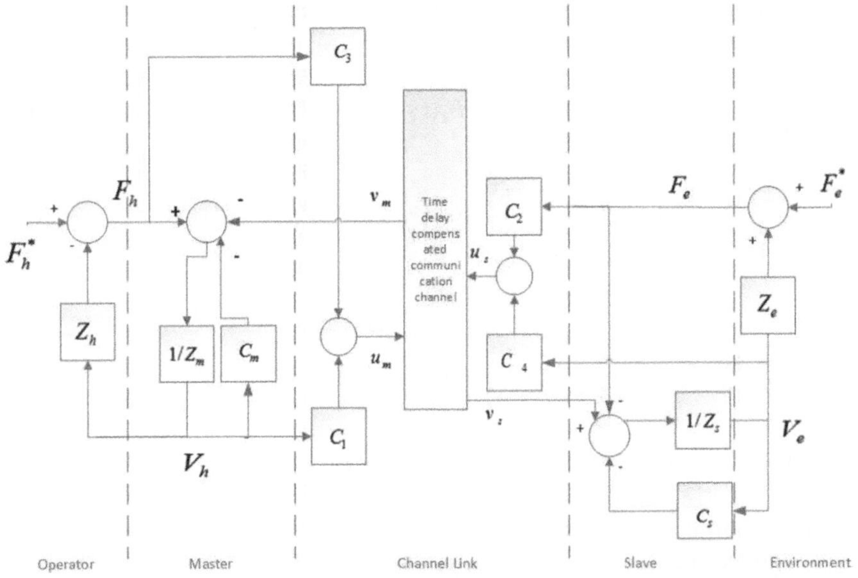

Figure 6.13. *4CH telerobotic system using wave theory for delay compensation (Aziminejad* et al. *2008). The central part (time-delay-compensated communication channel) is detailed in Figure 6.14*

6.7. Other methods for time-delay mitigation

In the preceding sections, the control theory approach is used to mitigate transmission time delay. Recently, other methods, such as time series prediction techniques, have been proposed for both statistical and machine learning approaches (Farajiparvar *et al.* 2020). In time series prediction, the aim is to predict future values based on past observations. Methods such as auto-regressive (AR), moving average (MA), as well as auto-regressive-moving-average (ARMA) models (Hamilton 1994) are used to predict the time delay in the transmission channel. The prediction provides knowledge about the time delay in advance, which can then be substituted in equation [6.10] to adapt the control variables (C_i) for stability and transparency. The neural networks approach, such as the recurrent neural network (RNN), is used for time series prediction (Li *et al.* 2008).

The new machine learning approaches for time series prediction above may open a promising avenue for solving the time delay problem in teleoperation systems. Compensation of the predicted delay values can be used to compensate for the time delay.

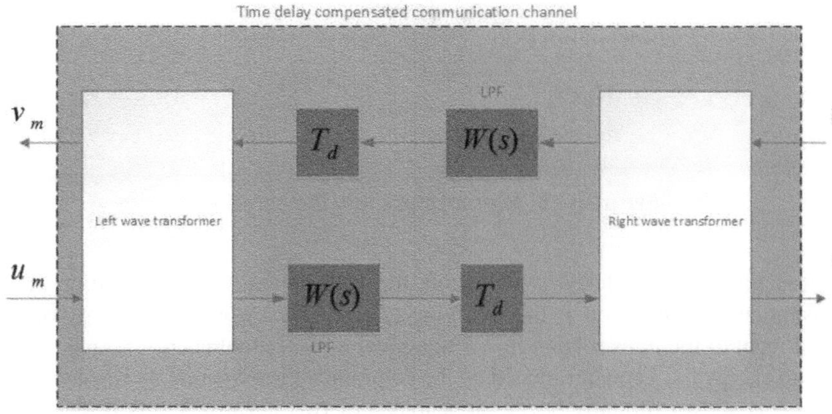

Figure 6.14. *Detailed block diagram of the time-delay-compensated communication channel shown in Figure 6.13*

6.8. Teleoperation over the Internet

Historically, the first teleoperation system working over the Internet was the "Mercury Project" in 1995 (Goldberg *et al.* 1995). Other projects followed for controlling underwater vehicles (Bruzzone *et al.* 2003). The technical difficulty of these systems is the variable transmission delays, since the packet data takes different paths along the Internet. As shown in Figure 6.15, the transmission delay is different in both directions and varies with time. This is also known as jitter, which is the difference in packet delay by measuring the difference in packet inter-arrival time.

The analysis of stability and passivity in the previous sections covered constant delays, in continuous time, over the transmission line. However, in practice, transmission delays vary. They are discrete, especially for a communication link over the Internet or the new generation of wireless communication systems based on IP protocols such as 4G and 5G. This is because of the transmission of data in small packets and their routing through different possible paths. The average latencies may be low, but instantaneous variations may appear due to the routing and increased network traffic.

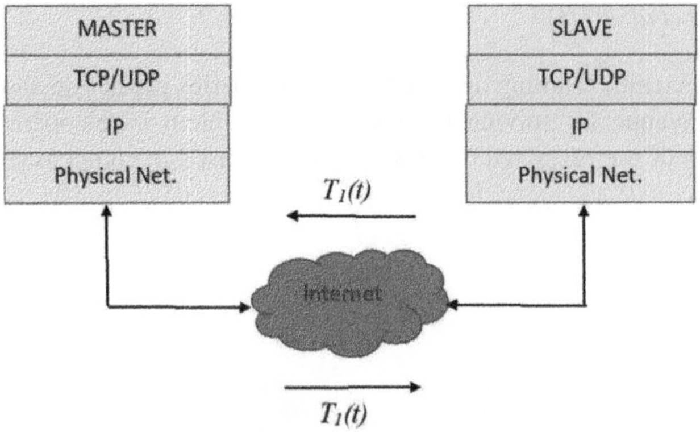

Figure 6.15. *Internet stack with the network*

These time-varying delays will distort the transmitted signal, resulting in its expansion and compression at the receiving end. In other words, the energy of the delayed input to the transmission line is not equal to the output energy, which means that extra energy has been introduced by the communication channel. In Lozano *et al.* (2002), it is shown formally by the appearance of two negative terms in the input energy equation, as in equation [6.12], which are not due to the stored energy but due to the input waves from the master and slave sides. As a result of this energy generation by the communication channel, instability can be introduced to the system. A natural approach to tackle this problem is by dissipating this energy.

In the same way as in equation [6.31] and using the variable delay network domain shown in Figure 6.16

$$u_s(t) = u_m(t - T_1(t)) \quad v_m(t) = v_s(t - T_2(t)) \qquad [6.39]$$

where $T_1(t)$ and $T_2(t)$ are the forward and feedback path delays, respectively. The estimation of the energy, as shown in equations [6.32] and [6.34], and as shown in (Lozano *et al.* 2002; Chopra *et al.* 2003), using the function f_i, which passivates the system, should be:

$$f_i^2 \leq 1 - \frac{dT_i}{dt} \quad i = 1, 2$$

The gain factors f_is will certainly preserve passivity, but, on the other hand, they will affect the performance of the system.

A list of control methods to obtain better teleoperation performances over the Internet, by adding adaptive and model predictive algorithms to the wave-variable technique above, is summarized in Kebria *et al.* (2019). Choosing the correct method between them depends on the requirement of the applications.

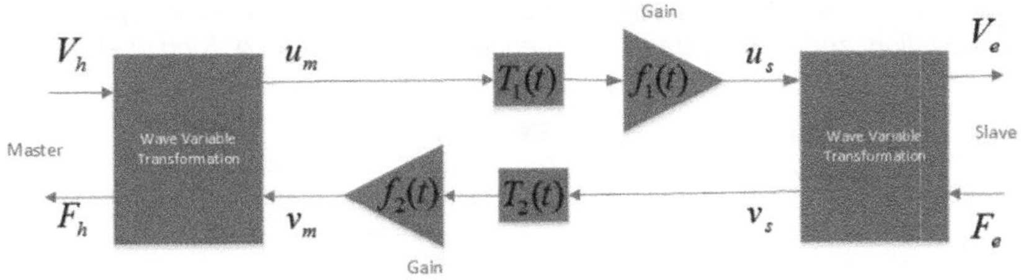

Figure 6.16. *Variable delay representation in the network domain*

6.9. Multiple access to a teleoperation system

Before the advent of the Tactile Internet, telepresence was used for audio/video teleconferencing, as in the ITU telecommunication standardization of telepresence system architecture (ITU:T 2014). An added value to this telepresence is added mobility to the interactive audio/video interfaces, using a controlled mobile robot. The robot can then move in a manner controlled by a master exploring the remote environment with audio and video feedback. This type of teleoperated mobile robot can foster social interaction between individuals (Kristoffersson *et al.* 2013).

Adding the haptic and movement capabilities to the audio, video and mobility capabilities adds more value to the telepresence system in order to reconstruct an authentic slave domain environment for the human operator on the master domain.

The architecture of such a system will be a mixture of a control system, multimodal communication channels and communication session management. The communication channel is mainly based on the IP protocol over the Internet. It can then connect different combinations of master–slave couples from multiple different users: from masters to slaves. A flexible framework for initiating, handling and terminating Internet-based telepresence sessions is necessary. The standard session initiation protocol (SIP) can establish the communication and then allow the multimodal data exchange. The parameter negotiation and transport protocol selection for the multimodal data is made. Finally, following the task termination, the session will also be terminated. Figures 6.17 and 6.18 show the sequence of all the steps from the establishment of a connection to the session ending (King *et al.* 2010).

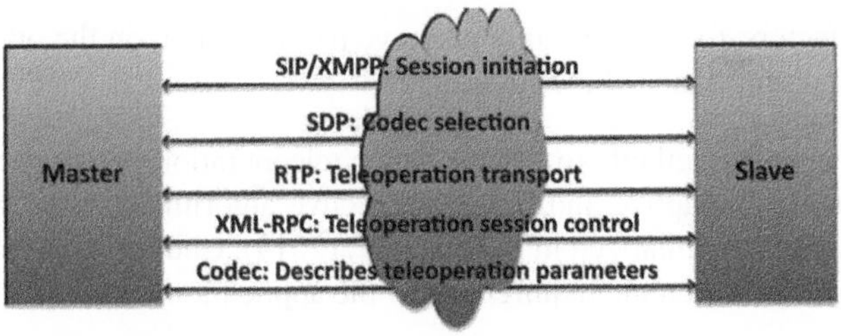

Figure 6.17. *Use of SIP for session management, SDP for codec and parameter negotiation, and RTP for transport of encoded media (source: King* et al. *2010)*

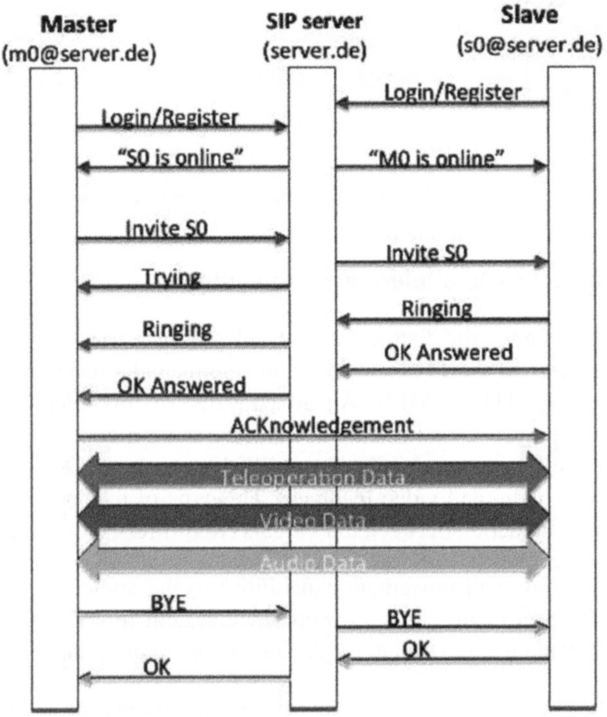

Figure 6.18. *SIP master to slave call with both users addressed on the same server (source: King* et al. *2010)*

TCP/IP is too heavy for telepresence applications, which would generate congestion in the network and, consequently, transmission delay. The delay varies with the congestion, bandwidth or distance the signal travels. This variable delay will create instability in the telepresence system. However, UDP/IP communications perform better in time delay aspects. However, packet loss is an issue. A buffering and interpolation method is suggested in order to solve the packet loss problem by Berestesky *et al.* (2004). In Arata *et al.* (2007), a trial of telesurgery was made, using UDP to decrease the transmission delay time. In the next chapter, we will present other possible protocols for faster transmission of packets over the Internet to solve the problem of packet loss and network congestion.

Figure 6.19. *Twenty actuated DOF and a further four under-actuated movements for a total of 24 joints in a dexterous hand from The Shadow Robot Company (source: https://www.shadowrobot.com/dexterous-hand-series/)*

Multilateral shared control architecture is possible; for example, a dual-user teleoperation system is suggested by Khademian and Hashtrudi-Zaad (2012). The proposed controller allows interaction between two users as well as the slave and environment through a dominance factor, which adjusts the authority of the users over the slave robot and environment. Therefore, the Internet is the simplest available infrastructure for such types of teleoperation systems with multiple operators.

6.10. A use case

With the commercial availability of slave robotic mechanisms and master analogous sensing mechanisms, such as the shadow and the Haptx glove, a complete tactile system can be realized. Figures 6.19 and 6.20 show these systems to be connected via a reliable communication networks to give haptic feedback to the glove while controlling the movements of the robotic hand. In section 9.4, a detailed possible generic robotic hand and glove is given.

The Shadow Robot Company has tested this system. A demonstration can be found in the videos (The Shadow Robot Company 2019a, 2019b).

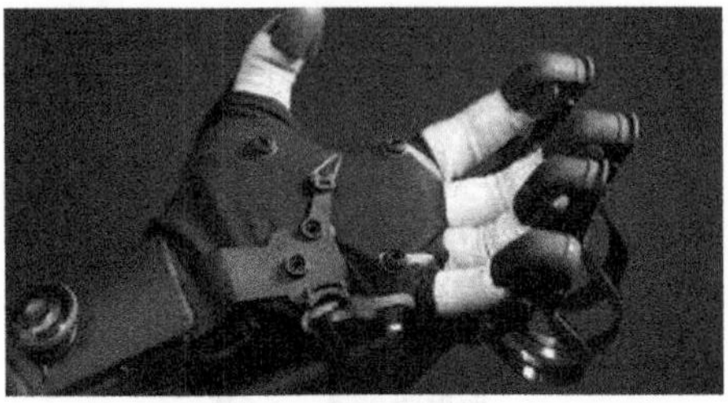

Figure 6.20. *Hapex glove contains haptic feedback. Connecting to the shadow robotic hand, it can mimic the glove movements (source: https://haptx.com/robotics/)*

6.11. Conclusion

In this chapter, we introduced a brief historical overview of teleoperation. Teleoperation's evolution to telepresence was possible thanks to the advances in communication channel characteristics, such as reliability, bandwidth and end-to-end delay. Analysis of the system improved the telepresence systems' stability and transparency by adding local control loops to correct the effects of the transmission medium. However, a compromise between transparency and stability is needed. For example, to realize control through a delayed transmission, we need to damp the system, but in doing so, we lose transparency, reducing the QoE. Prediction methods have also been used recently to solve the problem of transmission delay and its variations. Successful telepresence trials over the Internet have been made, and solutions for the packet loss and variable transmission delays have been suggested and are being applied.

6.12. References

Anderson, R.J. and Spong, M.W. (1989). Bilateral control of teleoperators with time delay. *IEEE Transactions on Automatic Control*, 34(5), 494–501.

Anvari, M., McKinley, C., Stein, H. (2005). Establishment of the world's first remote surgical service for provision of advanced laparoscopic surgery in a rural community. *Annals of Surgery*, 214(3), 460–464.

Arata, J., Takahashi, H., Pitakwatchara, P., Warisawa, S., Tanoue, K., Konishi, K., Ieiri, S., Shimizu, S., Nakashima, N., Okamura, K., Fujino, Y., Ueda, Y., Chotiwan, P., Mitsuishi, M., Hashizume, M. (2007). A remote surgery experiment between Japan and Thailand over internet using a low latency CODEC system. *Proceedings 2007 IEEE International Conference on Robotics and Automation*, 953–959, 10–14 April, Rome, Italy.

Aziminejad, A., Tavakoli, M., Patel, R.V., Moallem, M. (2008). Transparent time-delayed bilateral teleoperation using wave variables. *IEEE Transactions on Control Systems Technology*, 16(3), 548–555.

Bejczy, A.K. (1980). Sensors, controls, and man-machine interface for advanced teleoperation. *Science*, 208(4450), 1327–1335.

Berestesky, P., Chopra, N., Spong, M.W. (2004). Theory and experiments in bilateral teleoperation over the internet. *Proceedings of the 2004 IEEE International Conference on Control Applications, 2004*, 1, 456–463, 2–4 September, Taipei, Taiwan.

Bruzzone, G., Bono, R., Bruzzone, G., Caccia, M., Coletta, P., Spirandelli, E., Veruggio, G. (2003). Internet-based teleoperatlon of underwater vehicles. *IFAC Proceedings*, 36, 193–198.

Chopra, N., Spong, M.W., Hirche, S., Buss, M. (2003). Bilateral teleoperation over the internet: The time varying delay problem. *Proceedings of the 2003 American Control Conference, 2003*, 1, 155–160, 4–6 June, Denver, USA.

Colgate, J.E. and Brown, J.M. (1994). Factors affecting the Z-Width of a haptic display. *Proceedings of the IEEE 1994 International Conference on Robotics and Automation*, 3205–3210, 8–13 May, San Diego, USA.

Daniel, R.W. and McAree, P.R. (1998). Fundamental limits of performance for force reflecting teleoperation. *International Journal of Robotic Research*, 17(8), 811–830.

Dipietro, L., Sabatini, A.M., Dario, P. (2008). A survey of glove-based systems and their applications. *IEEE Transactions on Systems, Man, and Cybernetics, Part C (Applications and Reviews)*, 38(4), 461–482.

Eid, M., Cha, J., El Saddik, A. (2008). HugMe: A haptic videoconferencing system for interpersonal communication. *2008 IEEE Conference on Virtual Environments, Human-Computer Interfaces and Measurement Systems*, 5–9, 14–16 July, Istanbul, Turkey.

Farajiparvar, P., Ying, H., Pandya, A. (2020). A brief survey of telerobotic time delay mitigation. *Frontiers in Robotics and AI*, 7, 198 [Online]. Available at: https://www.frontiersin.org/article/10.3389/frobt.2020.578805.

Ferrell, W.R. and Sheridan, T.B. (1967). Supervisory control of remote manipulation. *IEEE Spectrum*, 4(10), 81–88.

Goldberg, K., Mascha, M., Gentner, S., Rothenberg, N., Sutter, C., Wiegley, J. (1995). Desktop teleoperation via the World Wide Web. *Proceedings of 1995 IEEE International Conference on Robotics and Automation*, 1, 654–659, 21–27 May, Nagoya, Japan.

Hamilton, J.D. (1994). *Time Series Analysis*, 1st edition. Princeton University Press, Princeton [Online]. Available at: http://gen.lib.rus.ec/book/index.php?md5=452d247f10f10f3d402e98b16383461b.

Hannaford, B. (1989). A design framework for teleoperator with kinesthetic feedback. *IEEE Transactions on Robotics and Automation*, 5(4), 426–434.

Heck, D. (2015). Delayed bilateral teleoperation: A direct force-reflecting control approach. PhD Thesis, Technische Universiteit Eindhoven, Eindhoven, Netherlands.

Higuchi, K., Fujii, K., Rekimoto, J. (2013). Flying head: A head-synchronization mechanism for flying telepresence. *23rd International Conference on Artificial Reality and Telexistence ICAT*, 28–34, 11–13 December, Tokyo, Japan.

ITU:T (2014). Telepresence system architecture. Recommendation, H Series: Audiovisual and multimedia systems, H.420. International Telecommunication Union.

Jury, E.I. (1964). *Theory and Application of the Z-Transform Method*. John Wiley & Sons, New York.

Kebria, P.M., Abdi, H., Dalvand, M.M., Khosravi, A., Nahavandi, S. (2019). Control methods for internet-based teleoperation systems: A review. *IEEE Transactions on Human-Machine Systems*, 49(1), 32–46.

Khademian, B. and Hashtrudi-Zaad, K. (2012). Dual-user teleoperation systems:New multilateral shared control architecture and kinesthetic performance measures. *IEEE/ASME Transactions on Mechatronics*, 17(5), 895–906.

King, H.H., Hannaford, B., Kammerl, J., Steinbach, E. (2010). Establishing multimodal telepresence sessions using the Session Initiation Protocol (SIP) and advanced haptic codecs. *2010 IEEE Haptics Symposium*, 321–325, 25–26 March, Waltham, USA.

Kristoffersson, A., Coradeschi, S., Loutfi, A. (2013). A review of mobile robotic telepresence. *Advances in Human-Computer Interaction*, 2013(902316) [Online]. Available at: https://doi.org/10.1155/2013/902316.

Lawrence, D.A. (1993). Stability and transparency in bilateral teleoperation. *IEEE Transactions on Robotics and Automation*, 9(5), 624–637.

Lei, X., Zhang, T., Chen, K., Zhang, J., Tian, Y., Fang, F., Chen, L. (2019). Psychophysics of wearable haptic tactile perception in a multisensory context. *Virtual Reality and Intelligent Hardware*, 1(2), 185–200.

Li, W., Luo, Y., Zhu, Q., Liu, J., Le, J. (2008). Applications of AR*-GRNN model for financial time series forecasting. *Neural Computing and Applications*, 17, 441–448.

Lozano, R., Chopra, N., Spong, M. (2002). Passivation of force reflecting bilateral teleoperators with time varying delay. *Proceedings of the 8th Mechatronics Forum International Conference*, 24–26, 24–26 June, Enschede, Netherlands.

Minsky, M. (1980). Telepresence. *Omni*, 45–51.

Niemeyer, G. and Slotine, J.E. (1991). Stable adaptive teleoperation. *IEEE Journal of Oceanic Engineering*, 16(1), 152–162.

Sheridan, T.B. (1989). Telerobotics. *Automatica*, 25(4), 487–507.

Sheridan, T.B. (1992). *Telerobotics, Automation, and Human Supervisory Control.* The MIT Press, Cambridge.

Sheridan, T. (1995). Teleoperation, telerobotics and telepresence: A progress report. *Control Engineering Practice*, 3(2), 205–214 [Online]. Available at: http://www.sciencedirect.com/science/article/pii/096706619400078U.

Skubic, M., Morgan, B., Graves, S., Kondraske, G.V., Khoury, G.J., Fiedler, P. (1995). Performance measurement and prediction in a distributed telerobotics system. *1995 IEEE International Conference on Systems, Man and Cybernetics. Intelligent Systems for the 21st Century*, 3, 2121–2126, 22–25 October, Vancouver, Canada.

Tachi, S. (1992). Tele-existence. *Journal of Robotics and Mechatronics*, 4(1), 7–12.

Tavakoli, M., Aziminejad, A., Patel, R.V., Moallem, M. (2008). Discrete-time bilateral teleoperation: Modelling and stability analysis. *IET Control Theory Applications*, 2(6), 496–512.

The Shadow Robot Company (2019a). TACTILE TELEROBOT – WIRED DEMO. Youtube video [Online]. Available at: https://www.youtube.com/watch?v=XdERNxRqQ4Q [Accessed 20 May 2021].

The Shadow Robot Company (2019b). Jeff Bezos, Amazon's CEO with our Tactile Telerobot. Youtube video [Online]. Available at: https://www.youtube.com/watch?v=ZGugadTuFCA [Accessed 20 May 2021].

Wang, D., Song, M., Naqash, A., Zheng, Y., Xu, W., Zhang, Y. (2019). Toward whole-hand kinesthetic feedback: A survey of force feedback gloves. *IEEE Transactions on Haptics*, 12(2), 189–204.

7

Haptic Data: Compression and Transmission Protocols

Wrya MONNET

Department of Computer Science and Engineering,
University of Kurdistan Hewlêr, Erbil, Iraq

7.1. Introduction

The architecture and the mathematical model of a remote-controlled robot were presented in Chapter 6. The primary returned signal used to control the remote robot was the force signal. This signal is, in fact, a haptic, and more specifically, a kinesthetic signal, which contains information about the remote environment for a better control command by the human operator. Tactile signals can also be returned to the human operator to improve the perception of the manipulated objects. The haptic signals then enhance the compliance of the remote robot with the environment and things.

The transmission of the haptic signals from the slave side to the master through the communication channel should have an acceptable quality. In this chapter, haptic signal characteristics are presented along with their perception by machines and humans. Understanding the perception by humans helps to find a way to reduce redundancy and compress them for an efficient transmission bandwidth use.

Haptic interface devices, which are force-reflecting interfaces, are the control tool used in telerobotics for a dexterous hand manipulation of the remote environment and objects. The working principles of some commercial haptic interface devices and some haptic sensors and actuators are presented.

Finally, some communication protocols are given for the transmission of such signals over the Internet. These should have near-real-time performance since the human perception of haptic signals is within 1 ms; hence, very low transmission delay is required to transmit such signals.

7.2. Haptic perception

Unlike other senses, there is no independent organ for humans to provide the sense of touch. The sense of touch is distributed all over the skin with haptic receptors to collect the information. Since most robots are a close imitation of humans, this sense should also be distributed on robots' contact areas with objects and environments. A brief presentation of the human haptic sensors is given in the following sections, based on which the robotic sensors are built. This is followed by an explanation of how the robots perceive and model the manipulated objects.

7.2.1. *Human haptic perception*

Human haptic receptors reside and are distributed in the skin to form the sense of touch. These receptors can detect pressure, heat, cold and pain (Schmidt 1986). Pressure sensors or mechanoreceptors are the ones responsible for haptic and tactile sensations. These receptors have an absolute and relative threshold value of detection. Concerning the absolute perception, this results from the deformation of the skin due to applied pressure. As such, indentations of the skin of the order of magnitude of 0.01 mm (10 μm) suffice to produce tactile sensations when they occur on the inner surface

$$\frac{\Delta I}{I} = k, \quad \text{or } \Delta I = kI \qquad [7.1]$$

where ΔI is the difference threshold (JND), I is the initial stimulus and k is the constant that signifies that the ratio stays constant. The above Weber equation states that the psychophysical perception of a signal change is therefore proportional to the stimulus intensity itself. In general, the JND for human haptic perception ranges from $k \leq 15\%$. If a change in haptic force magnitude is less than this range, the user would not perceive that force (Nitsch *et al.* 2010). The margin of no perception can be used for data compression, as we will see in section 7.4.

7.2.2. *Telerobotic tactile and haptic perception*

The main control operation of a teleoperated robot in the Tactile Internet comes from the operator at the master domain. The role of sensing is to detect information before encoding it and conveying it by a communication channel for reproduction on the teleoperator side. The information consists of the status of the environment surrounding the slave robot. The teleoperation is then composed of a fast continuous, uninterrupted sensing, encoding and reproduction of tactile sensations. However, intermittent autonomous behavior may also be required on the slave side to encounter any extended delays in the sensory signals' continuous feedback loop to the master operator. This is similar to a human who acquires sensory-motor coordination through learning based on a considerable amount of sensed data accumulated in past practices, as shown by Arimoto *et al.* (2005); Inoue and Hirai (2014), who show the analogy of this human capacity on the two-finger robotic hand. In these works, a model and a proportional–integrative–derivative (PID) control system are used to grasp objects without the need for tactile sensing on the robotic fingertips. However, most grasping robotic hand systems depend on tactile sensing for the controller feedback. In such cases, perception of the object's material and shape is essential for a local assistance of the master-slave controlled robots.

A perception process is required to interpret and represent touch sensing information to observe and determine the object properties and a model for eventual local decision-making. It is essential in the absence or interruption of master control signals to let the telerobotic recognize unintentional collisions or make intentional physical contact with objects or humans, or in other words to improve environmental awareness when dealing with objects and humans.

The detection and perception of the materials and objects by a telerobot is not only useful for autonomously executed tasks and local decisions, but it can also be used in the compression of feedback signals used for the master domain reconstruction since realistic sensing is required for the distant human operator with minimum bandwidth utilization. As a result of the perception, some packed parameters extracted from a model can be fed back to the master to reconstruct the haptic effects on the master side. In addition to models, machine learning can also be used by the robot tactile sensing to recognize objects according to their properties and materials, which can be done to a very good extent (Xie *et al.* 2019).

For authentic feedback, the relevant tactile information for TI or telepresence applications extracted from sensing data are the shape, material properties and the object pose (Luo *et al.* 2017). In the following sections, the different techniques for these types of sensing are covered.

7.2.3. *Tactile sensing for material recognition*

The functioning of the Tactile Internet is comparable to a prosthetic hand device with tactile sensing feedback used to improve the grasp task's success rate. Therefore, an object's surface material properties are essential for an interactive action with the surrounding environment. The most critical parameters to be detected and conveyed by a slave robot to the human master are the surface texture and object stiffness. Tactile sensors detect the object's surface by a sliding exploratory movement on their surfaces, while the object stiffness can be detected using force sensors. Concerning the texture, many types of sensors with their own data processing methods are available, such as acoustic sensors with a frequency domain analysis of the sensed data to detect the different textures. The stiffness, which is the inverse of compliance, shows the rigidity of the object. Compliance can be described as the relationship between an object's deformation and the force applied to it. Therefore, one method to estimate the compliance is by measuring a given skin-like sensor's deformation when using a constant force on it. Piezoresistive sensors with a linear variation of conductance as a function of the applied force, as shown in Figure 7.1, can be used in this case.

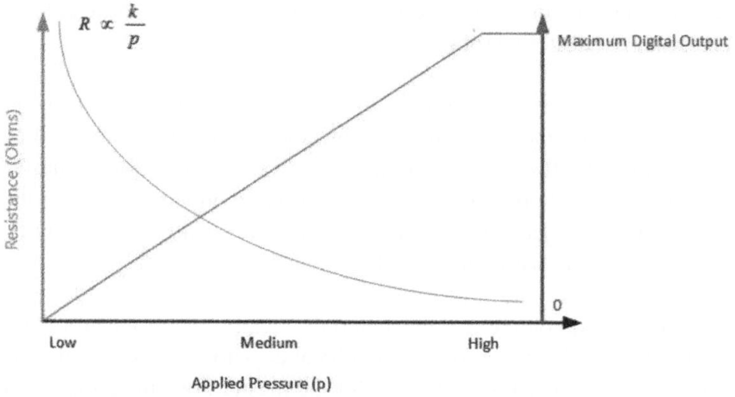

Figure 7.1. *Piezoelectric resistance characteristic with applied pressure. For a color version of this figure, see www.iste.co.uk/ali-yahiya/tactile.zip*

7.2.4. *Tactile sensing for object shape recognition*

The telepresence system assistance for the master side can be through identifying the shape of tackled objects. This can be for the aims of reconstruction, local control of objects or feature extraction for data compression. An object's perception can be done using image processing and recognition techniques invariant with image scale, translation and rotation (Lowe 1999). However, this may not be very practical in the control tasks of telerobotic and telepresence applications. In the latter cases, tactile object perception is necessary. Two tactile senses will interact to perceive the objects: the cutaneous sense to recognize the local shape and the combination of cutaneous and kinesthetic senses to detect the object's global contours. This can be summarized in the perception of a Rubik's cube and a simple smooth cube shown in Figure 7.2.

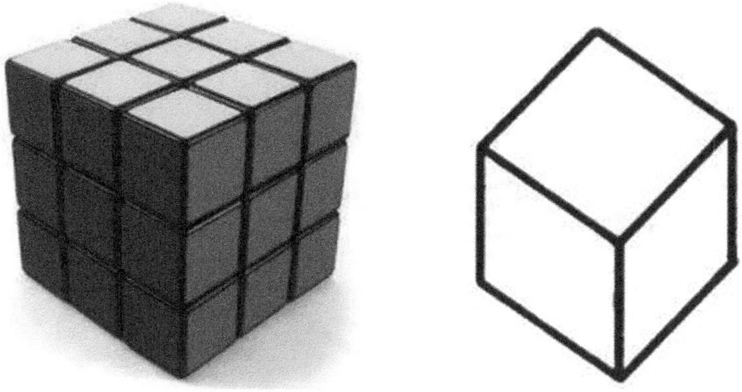

(a) (b)

Figure 7.2. *Two cubes: (a) is different in its local shapes from (b) which is smooth. Both are globally a cube. For a color version of this figure, see www.iste.co.uk/ali-yahiya/tactile.zip*

The sensors for this type of perception are mostly a two-dimensional array of sensors to detect the different amounts of applied force on the object. Figure 7.3 shows a tactile sensor organized in a matrix array of pressure sensors. Many data processing methods exist to recognize the local shape, such as raw data manipulation or statistical feature extraction. Otherwise, the data collected from the array of sensors can be seen as a two-dimensional image, with each sensor representing a taxel.

The local object recognition is then done by features adapted from computer vision applied to the array taxels such as moment estimation, feature extraction from the tactile image histogram, principal component analysis (PCA) and the data extract features. Finally, neural networks can be applied with unsupervised learning and incremental learning to classify objects. The latter method is also useful for global shape perception.

To estimate the global shape, the point cloud method is also used.

7.2.5. *Tactile sensing for pose estimation*

The object pose is not essential for Tactile Internet applications since vision assistance is also available for the master side. This helps with the localization and manipulation of objects with telerobotic arms and hands.

In summary, Table 7.1 shows the computational techniques used in object recognition using different tactile sensors and application domains. The table is extracted from a detailed review of the tactile sensing techniques of robot hands given in Kappassov *et al.* (2015).

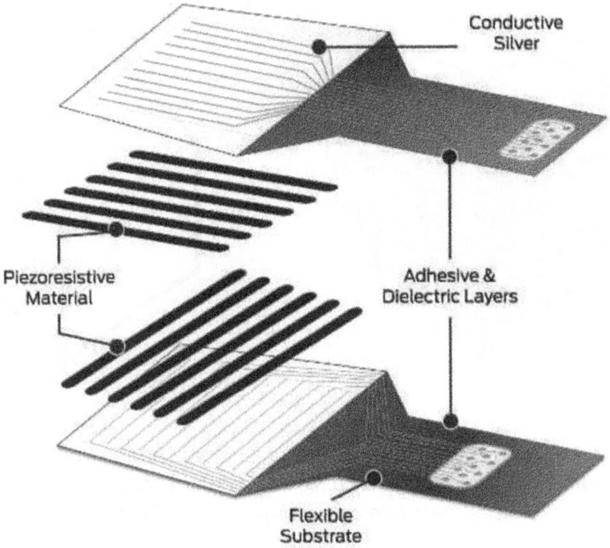

Figure 7.3. *Thin tactile sensor technology from (https://www.tekscan.com/). For a color version of this figure, see www.iste.co.uk/ali-yahiya/tactile.zip*

Tactile data type	Computational technique	Applications
– Force sensor – Vibrations – Tactile contact patterns	– Preprocessing (low-pass and high-pass filtering) – Fourier transform, STFT and FFT – Feature extraction: principle component analysis PCA, spectral power – Discrete wavelet transform – Spatial filtering – Classification: artificial neural network.	– Slip detection – Grip force control – Slip-type recognition – Texture recognition – Contact shape recognition

Table 7.1. *Overview of computational techniques applied to tactile sensing signals in the reviewed robot hand applications (Kokkonis* et al. *2018)*

7.3. Haptic interfaces

A haptic interface consists of a mechanism, sensors, actuators and hardware to estimate the human movement and return the forces from the slave. The hardware controls the mechanism through its impedance. The mechanical impedance gives a close perception of the remote object or environment. The sensors provide the position data, which can be derived once or twice to provide velocity and acceleration information.

7.3.1. *Haptic interface for telepresence*

Haptic perception refers to both the cutaneous (tactile) and kinesthetic sense, which conveys important information about distal objects and events (Loomis and Lederman 1986). Haptic perception plays a significant role in telerobotics to enable an effective human–machine interface in addition to video and audio interfaces. These interfaces are also used to manipulate virtual objects, in CAD/CAM applications, before manufacturing them. Artificial sensors and actuators are used to capture and convey this information to remote sites. For example, in the bilateral telepresence system, both the master and slave domains are equipped with these sensors and actuators. On the master domain side, as explained before, the ensemble of the sensors and actuators is called the haptic interface. It is used to translate the operator's manual dexterity while giving a realistic force reflection, position and touch feeling from the manipulator (slave domain).

Haptic interfaces enable human–machine communication through kinesthetic and tactile senses. Many anatomical parts of the human body, such as hands, skin and limbs, can provide the haptic communication channel between human and machine. However, the primary currently available commercial haptic interfaces are based on manual touch. A haptic exoskeleton is another way to communicate with a device. Commercial exoskeleton products for health and work, also called wearable robots, are available (https://exoskeletonreport.com/). These products are mainly known for personal and local uses other than for remote-controlled parts. However, in space research, more specifically the European Space Agency (ESA), a demonstration example of telepresence with exoskeleton can be found in Schiele (2014) and ESA (2014).

The haptic interface hardware design is based on a human contact scenario using hands, limbs or skin. It will also depend on the target application because of the complexity required to interface the human haptic sensory and mechanical movement. For example, suppose we need a plier gripper effector to execute a task at the slave domain in a telerobotic application. In that case, it is not necessary to use an exoskeleton-type haptic interface on the master side. A haptic interface's hardware can vary from a gaming joystick, a multiple DOF stylus to a wearable exoskeleton device

(Malley and Gupta 2008), for the kinesthetic to an array of factors for tactile sensing. Following the studies of Millman and Colgate (1991), Adelstein and Rosen (1992) and Minsky (1995) in force-reflecting and tactile interfaces, respectively, commercial haptic devices are now increasingly available. Figures 7.4 and 7.5 show some of the commercial haptic interface products available. Both haptic and tactile interfaces can be integrated into a single interface to feedback force and tactile from the remote device.

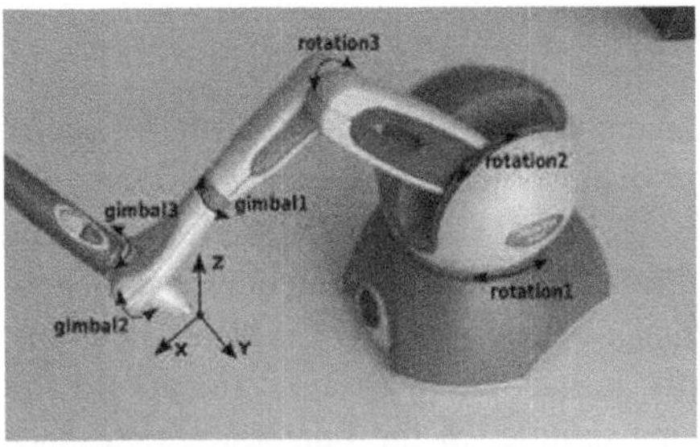

(a) Phantom Omni

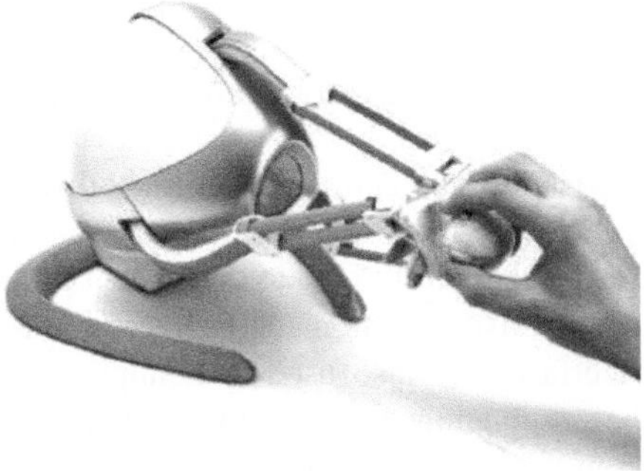

(b) Novint Technologies Falcon

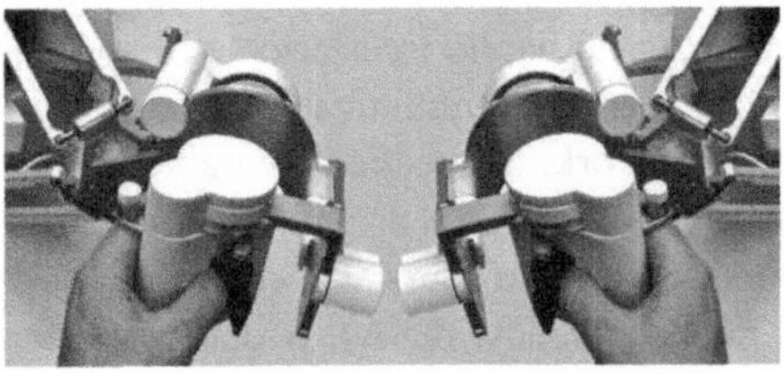

(c) Force Dimension Sigma 7

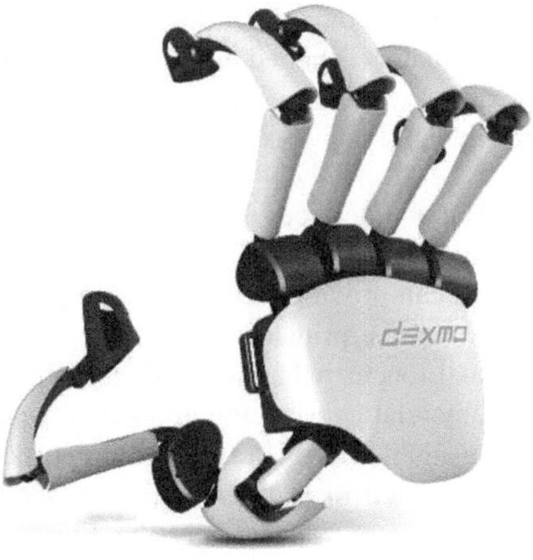

(d) Dexmo

Figure 7.4. *Some commercially available haptic interface devices. For a color version of this figure, see www.iste.co.uk/ali-yahiya/tactile.zip*

In 1980, Bejczy (1980) listed the difficulties in the mechanical design of haptic interfaces due to several problems, "These problems concern linkage of joint geometry, kinematic redundancy and dexterity, structural stiffness and dynamics, actuators for power and precision, motion transmission within the manipulator, mathematics for geometry and dynamics of mechanical arms, properties of end effectors and mechanical evaluation criteria.". He then listed the challenges in the human–machine interface research domain. Here are some that are still valid:

– the form of the sensors should be adapted to the real-world environment (task);

– the control and command language should be tailored to the mechanical, sensing and electronic properties of the manipulator system;

– the interface with the operator should be such that to intensify his or her capabilities by using increased degree of freedom (DOF), 3- to 6-DOF devices are available commercially.

Figure 7.5. *A tactile interface device: Lumen (Parkes* et al. *2008). For a color version of this figure, see www.iste.co.uk/ali-yahiya/tactile.zip*

To conclude this section, we cannot neglect the importance of video and audio information in telepresence applications. In fact, in industrial robotics, where automation plays the leading role, visual technologies are used by robots to outline objects. However, this is not sufficient; a more detailed and precise definition of objects and their handling are obtained using force-reflective sensing. In his visionary article, Sutherland (1965) argued that there is no reason only hands and arms should be used to control the computer due to their high dexterity. He adds that eye dexterity is also high and that machines to interpret eye motion will also be built. The range of interface devices, including haptic ones, can still increase based on new sensor and actuator technologies. In the following section, a list of haptic sensors and actuators is given as an example.

7.3.2. *Haptic and tactile sensors and actuators*

Tactile sensors are defined according to Lee and Nicholls (1999) as: "a device or system that can measure a given property of an object or contact event through physical contact between the sensor and the object". The property of the object includes its shape and texture. Many authors provide reviews on the types of sensors and their working principles (Rossi 1991; Amato *et al.* 2013; Dahiya and Valle 2013). This section presents the different kinds of tactile and kinesthetic sensing devices and a brief about the basics of their functioning.

– Resistive and piezoresistive sensors: the resistance of the sensor changes according to two variables; first, the contact position, and second, the contact force. The first variable is of the potentiometer type, and the second is of the piezoresistance type. In the potentiometer type, two resistive sheets are stacked in two layers. Each sheet is made of many lines of resistive voltage dividers, one horizontally and the other vertically. When stacked on each other, the two layers form a grid of resister circuits. When a point is pressed on the sheet, contact is made between the horizontal and vertical voltage dividers. The voltage dividers' output voltage in x and y directions is a function of the position of the contact. In the same way as the resistive sensor, to add a contact force sensing capability, two-layer lines of conductors are stacked with a piezoresistance material between two layers (Robertson and Walkden 1985). A pressure exerted on the surface will create a variable resistance that can be scanned with a particular electronic circuit to locate the position and evaluate the force as shown in Figure 7.6. Another technique that allows measurement of multiple hybrid resistive tactile sensing is given by Hong Zhang and So (2002). Another type of sensor based on piezoresistive touch sensors is also given in Goger *et al.* (2009).

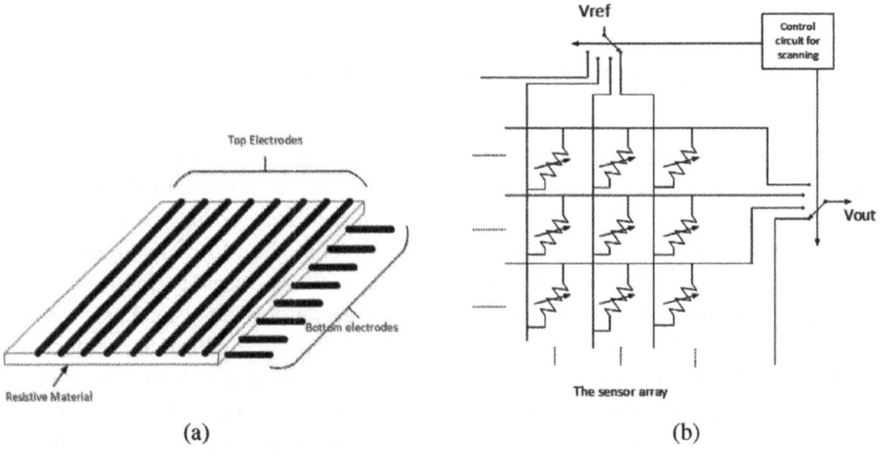

Figure 7.6. *Working principle of a piezoresistive touch sensor (Robertson and Walkden 1985): (a) construction; (b) components and interfacing. For a color version of this figure, see www.iste.co.uk/ali-yahiya/tactile.zip*

– Capacitive sensors: capacitive measurement methods are widely used nowadays in human–computer interfaces such as touch screens and trackpads. The functioning principles are based on the capacitance variation, which is a function of the distance between the plates and the area of its plates as shown in Figure 7.7. When a force is applied to the capacitor where the normal force component changes the plates' distance, the tangential force component changes the plates' effective area. This change in capacitance is then converted into a voltage change.

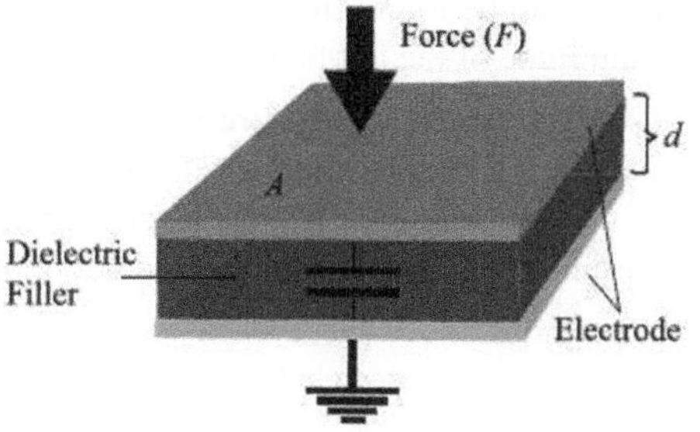

Figure 7.7. *Parallel plate capacitor consisting of two parallel plates of area A separated by distance d (Dahiya and Valle 2013). For a color version of this figure, see www.iste.co.uk/ali-yahiya/tactile.zip*

– Optical sensors: one example of this type of sensor is based on the refraction in an optical waveguide plate (Ohka 2007) as shown in Figure 7.8. When a light beam is directed into the plate, it remains within it unless the surrounding refractive index is higher than the optical waveguide. This happens when the rubber sheet, featuring an array of conical feelers, is pressed. The canonical feelers are deformed on the plate surface, which reflects the light totally. The distribution of contact pressure is calculated from the bright areas viewed from the plate's reverse surface. Another technique based on fiber Bragg grating (FBG) is given in Heo *et al.* (2006). Its working principle is based on the shifts in the reflected Bragg signal's wavelength as a function of external parameters such as strain or force.

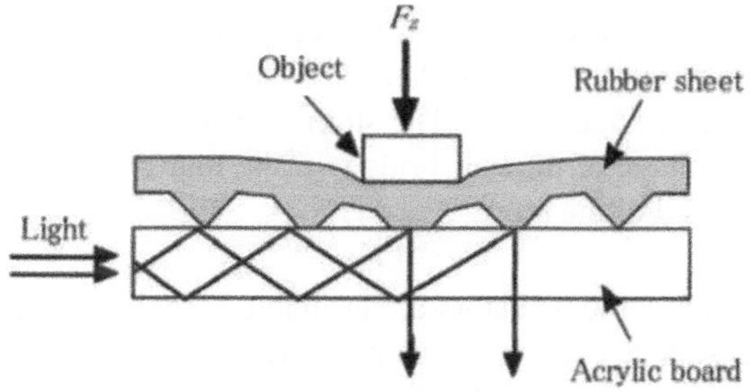

Figure 7.8. *Principle of optical tactile sensor (Ohka 2007). For a color version of this figure, see www.iste.co.uk/ali-yahiya/tactile.zip*

– Ultrasonic sensors: this involves a medium for ultrasonic signals composed of thin rubber. An ultrasonic transmitter and receiver are placed underneath this rubber covering. When an object, such as a finger, presses into the rubber, it is compressed and the path is reduced. By estimating this path and comparing it with the path of other regions, the difference can be found from the pressed area. The amount of path length depends on the magnitude of the applied force (Chuang *et al.* 2018).

– Magnetism-based sensors: these sensors are based on either the Hall effect or the inductance change of a coil. In Torres-Jara *et al.* (2006), the Hall effect is used to measure the deformation of a silicon dome when a force is applied to it. At the dome base, magnetic sensors are used to measure the deformation, which is a function of the applied force, as shown in Figure 7.9.

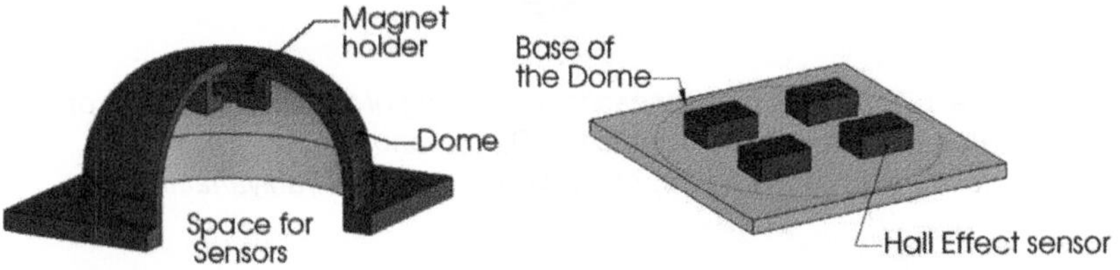

(a) The silicon dome (b) Circuit board with Hall effect sensors

Figure 7.9. *Magnetic touch sensor based on Hall effect (Torres-Jara* et al. *2006)*

– Microelectromechanical system (MEMS) sensors are realized by silicon micromachining, which integrates the micromechanical components and electronics on the same substrate. These devices are quite sensitive and result in higher spatial resolution. In Kane *et al.* (2000), the authors suggest a structure consisting of 4,096 elements arranged in an array of sensors on a silicon substrate. Each element is composed of a central shuttle plate suspended by four bridges over an etched pit. Each of the four bridges has a piezoresistor embedded in it. The structure's suspension over the substrate allows a deformation to occur in the bridges when normal and shear loads are applied to the central plate. The piezoresistors in each leg acts as the variable resistive half-bridge circuit. The strain can be measured by monitoring the intermediate node voltage of the half-bridge for the four bridges. Figure 7.10 shows the structure of one cell of the array.

Different types of tactile actuators are available based on some of the principles shown in the sensors above, such as piezoelectric actuators. In this case, the piezoelectric actuator converts the electrical energy into mechanical energy. Magnetic, electroactive polymer actuators and shape memory alloy actuators are other tactile actuators, as shown in Amato *et al.* (2013) and Lucia *et al.* (2019).

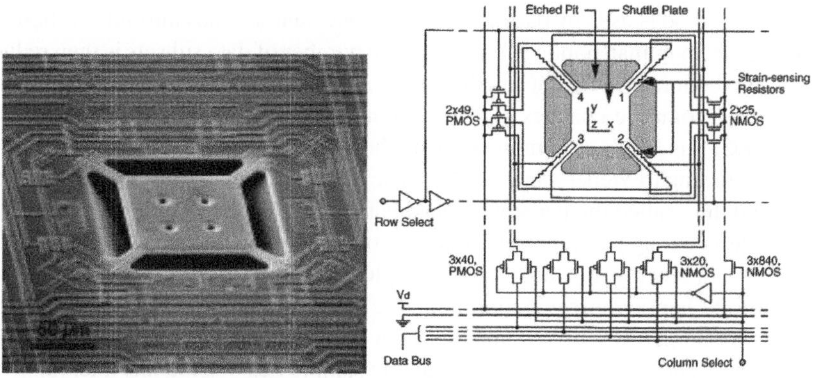

(a) The substrate (b) The schematic representation

Figure 7.10. *Single traction stress sensor consisting of a suspended plate/bridge structure with four embedded polysilicon resistors (Kane* et al. *2000)*

7.4. Haptic compression

The haptic data size can be reduced (compressed), like other types of data, such as multimedia content, for narrower transmission bandwidth. For example, lossless compression techniques such as Huffman can be applied, where the size of the haptic data is reduced without loss of information. However, this method is not efficient with regard to compression rate compared with the lossy compression algorithms. A subclass of the lossy method is based on humans' perceptual capabilities, such as MP3 for audio or MPEG4 for video content compression. In these algorithms, not only the redundant information but also the irrelevant information is removed. For example, in the case of MP3, some information that cannot be perceived or only weakly perceived by the human ears is removed based on the human audio perception model. Lossy perceptual compression for haptic signals can also be applied in the same way as audio signals, based on a model of human haptic perception and exploits its limitations.

One constraint in haptic data compression is in the bidirectional telerobotic control system. The signal processing algorithm to implement the compression should be done in real time within <1 ms since the global control closed loop's stability will be affected. This is in contrast to remote audio–video teleconferencing conversations, where the communications are twice unidirectional. Hence, audio and video signals can be compressed and sent without the need for the speaker, at one end, to listen to himself on the other end to control it, as shown in Figure 7.11.

Low-delay compression methods using differential pulse code modulation (DPCM) and adaptive DPCM with fixed and adaptive sampling frequency are presented and compared (Shahabi *et al.* 2002). In Guo *et al.* (2014), a linear prediction method is used by partitioning the haptic data samples into subsets based on knowledge from human haptic perception. The number of data subsets is then reduced using a prediction model. The predicted signal is derived such that it adapts itself to the local geometric changes of haptic signals. In their original work, Hinterseer *et al.* (2005) suggested a compression method based on the haptic discrimination bounds according to Weber's law of JND. The haptic information is not sensed in this bound, and thus called the perceptual dead band (PD). Therefore, the transmission of one sample is enough to represent a packet of samples within the PD, as shown in Figure 7.12. In other words, once a new sample is measured and exceeds the threshold value ε of the haptic discrimination bound, a packet containing the latest measured haptic information h is sent. Around this value, a new threshold interval $[h-\varepsilon, h+\varepsilon]$ is established, and only if a consecutive haptic sample lies outside this interval, a new packet is sent.

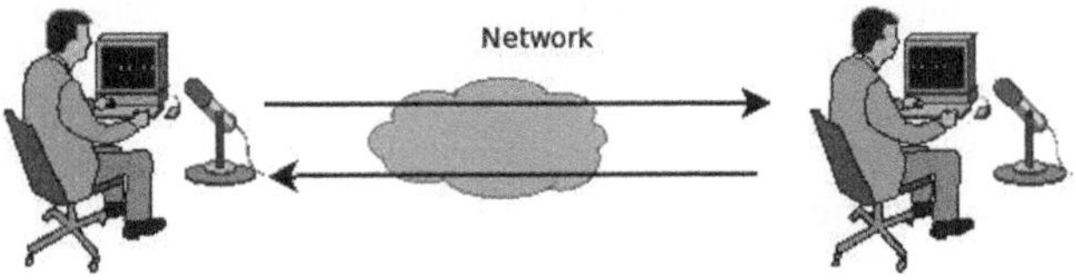

Figure 7.11. *Conversation in a video teleconferencing is two times unidirectional. For a color version of this figure, see www.iste.co.uk/ali-yahiya/tactile.zip*

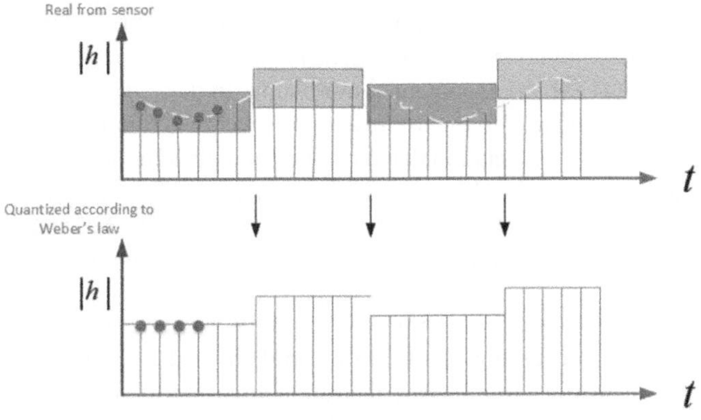

Figure 7.12. *Perceptual deadband compression. For a color version of this figure, see www.iste.co.uk/ali-yahiya/tactile.zip*

This algorithm can be used for different haptic sensor data such as position, velocity and force.

To apply the PD method above in a multi-DoF telerobotic system, h_i will become the new haptic sample to be compared with the previous in the Euclidean sense. If the distance $\|p_i - h_i\|_2 > \|p_i\|_2.k$, then the new sample vector h_i is transmitted. Otherwise, the new sample is discarded from the transmission (Kammerl 2012) as shown in Figure 7.13

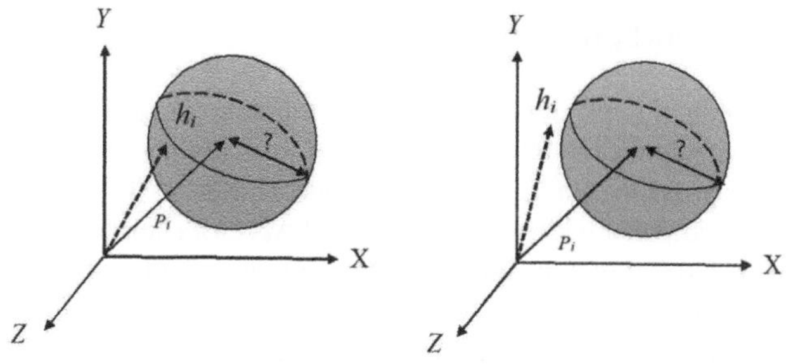

Figure 7.13. *Multi-DoF isotropic perceptual dead band PD ($\Delta = \|p_i\|_2.k$), from (Kammerl 2012). For a color version of this figure, see www.iste.co.uk/ali-yahiya/tactile.zip*

Additional methods can be used to predict new samples other than the simple sampling and hold described above. A first-order prediction can be used by taking two consecutive samples and finding a slope to be used in the prediction $p_i = h_j + c_j(i-j)$, where c_j is the slope of the line connecting the ith and j^{th} samples, with $i > j$.

The value of the parameter k is an essential one in the quality of the compressed signal. As given in section 7.2.1, its range can vary a lot according to different research. Optimally, its value should be high enough for a better compression ratio and low enough for a better haptic sensation and consequently a transparent teleoperation system.

Other studies (Lee and Payandeh 2012) have suggested adaptive quantization using signal processing methods, and in the study by Mizuochi and Ohnishi (2013), a coding scheme using a low-pass filter and discrete Fourier transform was applied with reduced calculation cost and packet size.

The above-mentioned methods are convenient for transmitting haptic information for real-time applications such as telerobotics and telepresence. In some haptic applications where haptic data is recorded for a virtual environment, the compression algorithm speed is not an issue, for example, algorithms using discrete cosine transform (DCT) (Nakano *et al.* 2015) and model-based compression algorithms (Hinterseer *et al.* 2006).

The following section will find some works where the compression algorithm is used in the application layer of transmission protocols. This compression technique also uses Weber's perception rule to sample higher or lower. This is the adaptive sampling in haptic applications, as suggested by Shahabi *et al.* (2001) and Otanez *et al.* (2002).

7.5. Haptic transport protocols

The transmission of haptic signals for teleoperation applications has specific needs such as reduced transmission delay and high transmission packet rate. Teleoperation applications mainly use the Internet as the medium of transmission in the absence and impracticality of dedicated wide area networks (WAN) for such applications. In order to achieve a good QoS in haptic signal transmission over the Internet and thus a good quality of experience (QoE) in the teleoperation applications, adapted transmission protocols should be used. These protocols should respect some QoS criteria and requirements. Table 7.2 from Kokkonis *et al.* (2018) shows the condition which the transmission media should respect to maximize the QoE of a haptic application.

QoS	Haptic	Video	Audio	Graphics
Jitter (ms)	≤ 2	≤ 30	≤ 30	≤ 30
Delay (ms)	≤ 50	≤ 400	≤ 150	$\leq 100-300$
Packet loss (%)	≤ 10	≤ 1	≤ 1	≤ 10
Update rate (Hz)	≥ 1000	≥ 30	≥ 50	≥ 30
Packet size (byte)	64–128	$\leq$ MTU	160–320	192–5,000
Throughput (kbps)	512–1,024	25,000–40,000	64–128	45–1,200

Table 7.2. *QoS requirement for different media streams (Kokkonis* et al. *2018)*

To satisfy the requirement mentioned above, for haptic systems, the transmission system's transport layer plays an important role. The protocols should then meet some constraints: the protocol should prioritize the real-time interactive data in multiplexing with audio and video signals. The protocols can moreover prioritize some essential samples of haptic information over others. A minimum overhead of transport protocol will allow a higher transmission rate. Concerning the jitter in reception and the congestion in the network, the protocol should pause transmission until the congestion is solved. Multiplexing and synchronizing different media streams with the haptic stream is another essential requirement for the transport protocol to obtain a good QoS.

Several protocols with some of the majority of the above requirements are suggested in the literature. Some of them are placed at the application layer, and others in the transport layer (Kokkonis *et al.* 2012). In the following section, we will give details about the properties of each of them.

7.5.1. *Application layer protocols*

The protocols in the application layer have the objective of customizing haptic communication protocols. This makes the protocol more agile. For example, a multi-modal or other additional information encapsulation is possible at this layer.

– Application layer protocol for haptic networking (ALPHAN) (Osman *et al.* 2008): this protocol operates on top of the UDP since it does not impose any reliability or flow control schemes. The protocol supports the notion of key updates widely endorsed by most haptic communication transport layer protocols. This is done in the application layer by implementing a reliability mechanism applied to such updates, while regular updates remain unaffected. Figure 7.14 shows the header format of the protocol, with the frame field description in Table 7.3

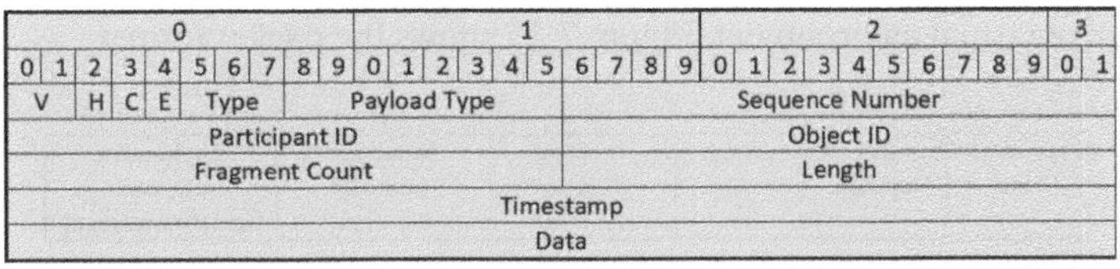

Figure 7.14. *ALPHAN header format from Osman* et al. *(2008)*

Fields	Bits	Description
V	2	The version of the protocol; set to 0 for the current version
H	1	This bit is set if the data is haptic, and is reset if the data is graphic
C	1	Reporting collision state: the payload has position and force information if 1, position only if 0
E	1	Not relevant for haptics. Indicate the last fragment of a packet for graphics data
Type	3	The type of haptic data: I, P, B frame and state query types for late comers
Payload type	8	Indicates single/multiple points of interaction for haptics and object types for graphics (rigid, deformable, etc.)
Sequence number	16	Identifies each packet and see if any packets were lost or delivered out of sequence
Participant ID	16	Identifies each participant in the C-HAVE environment.
Object ID	16	Identifies each object in the C-HAVE environment.
Fragment count	16	Not relevant for haptics. Identifies the number of fragments for graphic data
Length	16	Indicates the length of the payload
Timestamp	32	Time at which the event in the payload should be executed. Important for the implementation of a local lag mechanism

Table 7.3. *Frame field description (Osman* et al. *2008)*

Multiple buffering schemes are adapted in ALPHAN, where each haptic variable representing the environment is allocated a sending buffer. Two queues are arranged for each variable in case of retransmission to improve the application layer's reliability since the UDP is not providing it. Another advantage of buffering is in prioritizing variables to be sent according to their importance.

The I packets are sent reliably and are considered to be key packets. P packets are encoded differentially based on previous I or P packets, while B packets are encoded based on previous and future I and/or P packets.

– The synchronous collaboration transport protocol (SCTP) (Shirmohammadi and Georganas 2001): this is a host-to-host layer protocol, comparable to the transport layer. It is encapsulated into UDP packets and assumes that the underlying physical network supports IP multicasting. It was initially designed for interaction with a shared collaborative virtual environment. Figure 7.15 shows the packet format.

control bit	Key update	Object ID	Stream number	Sequence number	Application payload (update message)

Figure 7.15. *SCTP packet format*

The protocol adopts the concept of reliable packet delivery of key updates and unreliable delivery of regular updates. The object ID indicates which object is being interacted in the header; the key update specifies if the packet is a key update or not. The stream ID and sequence number can be used together to discard late arriving messages. The protocol is inherently time-sensitive since updates are immediately taken into consideration.

Another version of this protocol called the smoothed SCTP has been devised by Dodeller (2004) to overcome the receiver's jitter problem. A timestamp is added in the packet format. The time is divided into buckets of time of δt milliseconds at the receiver side. Each received update is put into a bucket according to its timestamp. A constant delay is added to the update; thus, all the update messages are processed with a fixed delay higher than the network delay, but constant.

– Perception-based adaptive haptic communication protocol (PAHCP) (Nasir and Khalil 2012). In this protocol, the author devised a modified version of the smoothed SCTP and took into consideration extra features from other protocols such as reliable transmission protocol (SRTP), reliable multicast transport protocol (RMTP), synchronous collaboration transport protocol (SCTP) and selective reliable multicast (SRM). With a fixed threshold variable, Weber's law is used to reduce the number of packets transmitted.

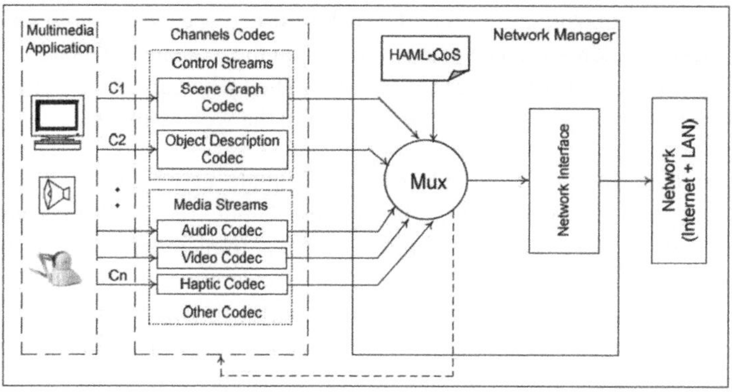

Figure 7.16. *Communication framework (Eid* et al. *2009)*

– ADMUX, Adaptive Multiplexer (Eid *et al.* 2011) is an adaptive application layer multiplexing framework including a communication protocol. It is designed for multimedia applications incorporating haptic, visual, auditory and scent data for non-dedicated networks. It uses statistical multiplexing to multiplex the different media. The statistical multiplexing principle is when a group of media channels shares a limited quantity of bandwidth. The allocation of bandwidth is on a frame-by-frame basis, which is controlled by the multiplexer. The multiplexer is based on a mathematical model that dynamically adapts to both the application needs and the network conditions (Eid *et al.* 2009). The haptic applications meta language (HAML) is used to initiate and functionally modify all the components of the communication framework as shown in Figure 7.16 (Eid *et al.* 2009). The components include:

- the transport protocol, which defines the transport layer protocol, its reliability mechanism and its QoS parameters;

- the synchronization scheme, which defines both the intermodal and intramodal synchronization models used in the communication;

- the compression scheme, which defines the data preprocessing method (if used) and the codec configurations;

- the control method scheme to describe the algorithms that are used at both ends of the network to compensate for the network deficiencies.

An advantage of the ADMUX is synchronizing haptic data with audio/video data and that it adapts to application level events and interactions.

– RTP/I real-time application level protocol for distributed interactive media (Mauve *et al.* 2001): initially, this was developed to implement distributive interactive media such as remote collaborative whiteboards and 3D design applications. The protocol reuses some of the RTP properties, such as using two protocols, one for data and the other for control, called RTP/I and RTCP/I, respectively, both carried over separate transport addresses. The protocol relies on the exchange of state information modified by either events or the passage of time. The technical details of this protocol are given in the document (Mauve *et al.* 2000) and the web resource[1].

– Haptic over Internet protocol (HoIP) (Gokhale *et al.* 2013, 2015). In this protocol, adaptive sampling (from the highest haptic rate 1,000 packets/s to a minimum of 10 packets/s) of the haptic variables is implemented based on Weber's law, where successive samples are compared to decide about the transmission or not of the new sample. HoIP uses the existing transport layer UDP for a better real-time performance and IP implementations. The frame of the HoIP protocol is shown in Figure 7.17. The fields of the frame are described in Table 7.4.

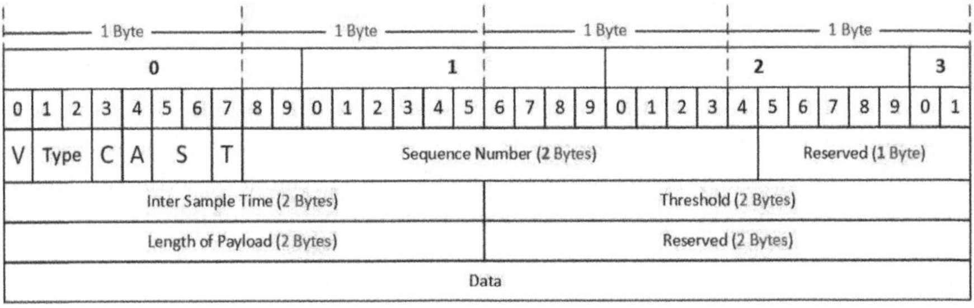

Figure 7.17. *The frame of HoIP protocol (Gokhale* et al. *2013). For a color version of this figure, see www.iste.co.uk/ali-yahiya/tactile.zip*

7.5.2. *Transport layer protocols*

– Interactive real-time protocol (IRTP): this is a transport layer protocol based on the IP; it is different from TCP and UDP but inherits their advantages (Ping *et al.* 2005). It is designed for interactive Internet-based service with a reduced end-to-end delay. It is a connection-oriented protocol, i.e. a connection must be established before transmitting data. This decreases its efficiency compared with connection-less protocols since in each connection some inquiry-reply happens before establishing a connection. But it has the advantage of managing the users and implementing reliable transmission since it is necessary. This protocol's short header is shown in Figure 7.18 with the fields' description given in Table 7.5.

The protocol interacts with the upper-layer protocol, in order to transmit both reliable and unreliable data. It distinguishes reliable transmission and unreliable transmission by the command segment in the header.

1. http://www.informatik.uni-mannheim.de/informatik/pi4/ projects/RTPI/index.html

Fields	Bits	Description
V	1	Protocol version set to 0 for haptic point-to-point communication (current implementation)
Type	2	Indicates type of the data. Set to 0 for haptics, 1 for haptics-audio, 2 for haptics-video, 3 for haptics-audio-video
C	1	Indicates content type of the payload. Set to 0 if only force coordinates are transmitted, 1 for position and velocity coordinates
A	1	Indicates the type of adaptive sampling employed. Set to 0 for Weber, 1 for level crossings. 1-bit is reserved for other samplers to evolve
S	2	Indicates transmission of threshold in the packet. Set to 1 if threshold is sent, else set to 0
T	1	Identifier of each packet to track lost or out-of-order packets
Sequence number	16	Indicates the inter sample time gap. Along with the received times, this can be used to calculate packet delays
Inter sample time	16	Value of the threshold in percentage. If set to 0, data is transmitted at the haptic loop rate
Threshold	16	Indicates length of the payload
Length of payload	16	Not relevant for haptics. Identify the number of fragments for graphic data

Table 7.4. *HoIP frame description (Gokhale* et al. *2013)*

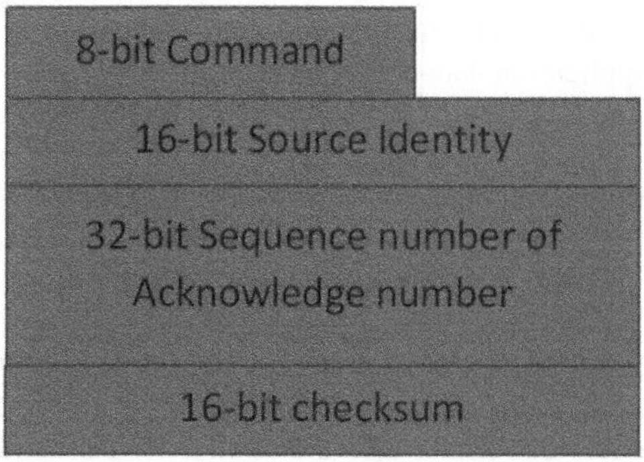

Figure 7.18. *IRTP protocol header consisting of nine bytes*

– Real-time protocol (RTP) (Schulzrinne *et al.* 1996) This provides end-to-end network transport functions suitable for applications transmitting real-time data, such as audio, video or simulation data, over multicast or unicast network services. It was primarily designed to satisfy the needs for multiple-participant multimedia conferences. RTP is carried on top of IP and UDP, where an RTP header precedes a chunk of data; both are in turn contained in a UDP packet. For example, in audio signal transmission, the RTP header indicates the type of audio encoding (PCM, ADPCM or LPC). The RTP header contains timing information to reconstruct the timing produced by the source.

Field	bits	Description
Command	8	4 groups of commands: connection related commands, data sending and receiving related commands, inquiry commands and timer-related commands
Source identity	16	To identify the local port of a connection while the remote Port uses the 16 bits of identification segment of the IP header.
Sequence number or acknowledge number	32	A 32-bit number is assigned to each byte of its data. The sender sets this segment as the sequence number of the first byte
Checksum	11 bytes	A 16-bit destination identity and 16-bit package length segment are added to the IRTP header and the checksum is calculated from it

Table 7.5. *IRTP header description*

The RTP is used in conjunction with the real-time control protocol (RTCP) and is also needed to establish a multiuser conference, where each instance of the audio, video or any other application data periodically multicasts a reception report plus the name of its user on the RTCP (control) port. The reception report indicates how well the current data is being received and may be used to control adaptive encodings. Other identifying information may also be included in this protocol subject to control bandwidth limits, such as the inter-arrival jitter estimate, information about the highest sequence number received, a fraction of packets lost and a cumulative number of packets lost.

– The stream control transmission protocol (SCTP) is a general-purpose transport layer protocol. The services it provides are similar to TCP and a set of advanced features (Stewart 2007) to use the enhanced capabilities of modern IP networks and support increased application requirements.

– The efficient transport protocol (ETP) (Wirz *et al.* 2008) is a transport layer protocol that optimizes the available bandwidth by sending the highest packets with a reduced round trip time (RTT) and the time between two packets' inter-packet gap (IPG). It is designed for haptic applications to integrate the good characteristics of many other transport protocols. The protocol's performance is based on reducing the IPG to reduce the RTT, which represents the frequency of closing the control loop of the remote haptic system $RTT = IGP.N$, where N is the sum of the number of sent and received packets through the network, ETP uses the UDP protocol for transporting its data.

– Real-time network protocol (RTNP) (Uchimura and Yakoh 2004): this is specially designed for bilateral teleoperation using master/slave manipulators and force feedback. It is specially designed for the UNIX environment. Its priority determines each packet order in a queue at the sender and the intermediate buffers over the network.

– Bidirectional transport protocol (BTP) provides a novel congestion control technique that enhances application and transport layer performance (Wirz *et al.* 2009). The protocol improves bilateral flow tasks for real-time telerobotics by minimizing the round trip time (RTT) while maximizing the transmission frequency. To do this, the network congestion is controlled by modifying the elapsed time between two consecutive packets sent: Inter-Packet Gap (IPG) . IPG is related to the interarrival time (IAT), which is the time elapsed between two consecutive arriving packets. Both are related and equal in the case of no congestion. The BTP packet information is shown in Figure 7.19, where it is shown that it uses the UDP protocol with an additional header of the BTP.

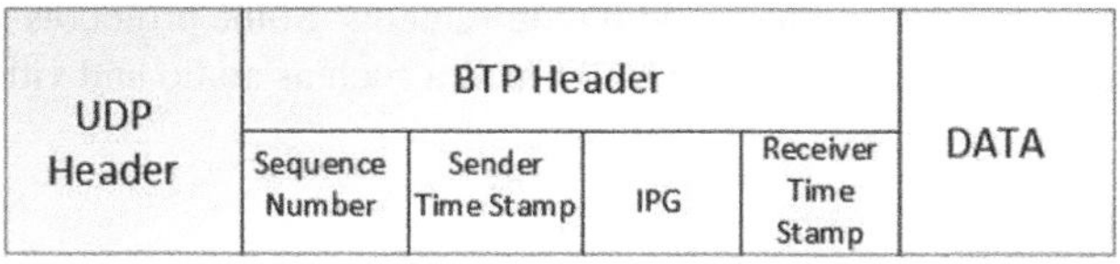

Figure 7.19. *BTP packet information*

7.6. Multi-transport protocols

Some protocols, as shown above, multiplex different media streams before transmitting them all in one protocol. Another way to send other data streams is to send them on multiple protocols according to their characteristics. For example, slow data with reliable communication requirements can be sent on a TCP protocol, while a control signal can be sent on a UDP and video or audio on an RTP. In Phung *et al.* (2012), the authors suggested the above multi-transport protocol in the remote control and monitoring of a mobile robot.

7.7. Haptic transport protocol performance metrics

The main performance metrics for haptic transport protocols are returned delay time, packet loss, jitter (varying time delay) and throughput. As was shown in Chapter 6, the constant delay in the network can be handled from the stability point of view. However, the variable delay causes system instability.

7.8. Conclusion

The compression and transmission of haptic data over a telepresence application network have different requirements and constraints from the other media-type signals such as audio and video. The samples of audio and video signals can be processed in packets and reduced in size using compression. The principle of perceptual predictive haptic data compression consists of two key components:

– a psychophysical model deployed at the encoder;

– signal predictors which are used at the encoder and the decoder.

The prediction algorithm estimates incoming haptic samples based on previously transmitted haptic data. Both predictors at the encoder and the decoder run in parallel and are fed with identical signal information to keep them strictly coherent. The psychophysical model of human haptic perception suitable for haptic real-time communication can be built based on Weber's law of JND. A signal prediction algorithm estimates future haptic samples based on previous signal input. For the prediction of haptic signals, typically linear predictors of a very low order and low latency are required.

Many protocols have been tried and suggested to transmit haptic information with its specific temporal and bandwidth requirements. Some protocols multiplex and synchronize haptic data with other types of media such as audio and video.

7.9. References

Adelstein, B.D. and Rosen, M. (1992). Design and implementation of a force reflecting manipulandum for manual control research. *ASME Winter Annual Meeting*, pp. 1–12.

Amato, M., Vittorio, M.D., Petroni, S. (2013). Advanced MEMS technologies for tactile sensing and actuation. *MEMS Fundamental Technology and Applications*. CRC Press, Boca Raton.

Arimoto, S., Ozawa, R., Yoshida, M. (2005). Two-dimensional stable blind grasping under the gravity effect. *Proceedings of the 2005 IEEE International Conference on Robotics and Automation*, pp. 1196–1202.

Bejczy, A.K. (1980). Sensors, controls, and man-machine interface for advanced teleoperation. *Science*. 208(4450), 1327–1335.

Chuang, C.H., Weng, H.K., Chen, J.W., Shaikh, M.O. (2018). Ultrasonic tactile sensor integrated with TFT array for force feedback and shape recognition. *Sensors and Actuators A: Physical*, 271, 348–355 [Online]. Available at: http://www.sciencedirect.com/science/article/pii/S0924424717317612.

Dahiya, R.S. and Valle, M. (2013). *Robotic Tactile Sensing Technologies and System*. Springer, Dordrecht.

Dodeller, S. (2004). Transport layer protocols for telehaptics update messages. *Proceedings of the 22nd Biennial Symposium on Communications.*

Eid, M., Cha, J., El Saddik, A. (2009). An adaptive multiplexer for multi-modal data communication. *2009 IEEE International Workshop on Haptic Audio Visual Environments and Games*, pp. 111–116.

Eid, M., Cha, J., El Saddik, A. (2011). Admux: An adaptive multiplexer for haptic audio visual data communication. *IEEE Transactions on Instrumentation and Measurement*, 60(1), 21–31.

ESA (2014). Exoskeleton to remote control robot live [Online]. Available at: https://www.esa.int/Science_Exploration/Human_and_Robotic_Exploration/Exoskeleton_to_remote-control_robot_live.

Goger, D., Gorges, N., Worn, H. (2009). Tactile sensing for an anthropomorphic robotic hand: Hardware and signal processing. *2009 IEEE International Conference on Robotics and Automation*, pp. 895–901.

Gokhale, V., Dabeer, O., Chaudhuri, S. (2013). Hoip: Haptics over internet protocol. *2013 IEEE International Symposium on Haptic Audio Visual Environments and Games (HAVE)*, pp. 45–50.

Gokhale, V., Chaudhuri, S., Dabeer, O. (2015). Hoip: A point-to-point haptic data communication protocol and its evaluation. *2015 Twenty First National Conference on Communications (NCC)*, pp. 1–6.

Guo, F., Zhang, C., He, Y. (2014). Haptic data compression based on a linear prediction model and quadratic curve reconstruction. *Journal of Software*, 9(11), 2796–2803.

Heo, J.S., Chung, J.H., Lee, J.J. (2006). Tactile sensor arrays using fiber bragg grating sensors. *Sensors and Actuators A: Physical*, 126(2), 312–327 [Online]. Available at: http://www.sciencedirect.com/science/article/pii/S0924424705006084.

Hinterseer, P., Steinbach, E., Hirche, S., Buss, M. (2005). A novel, psychophysically motivated transmission approach for haptic data streams in telepresence and teleaction systems. *Proceedings (ICASSP '05). IEEE International Conference on Acoustics, Speech, and Signal Processing*, 2, ii/1097–ii/1100.

Hinterseer, P., Steibach, E., Chaudhuri, S. (2006). Model based data compression for 3D virtual haptic teleinteraction. *2006 Digest of Technical Papers International Conference on Consumer Electronics*, pp. 23–24.

Inoue, T. and Hirai, S. (2014). Why humans can manipulate objects despite a time delay in the nervous system. In *The Human Hand as an Inspiration for Robot Hand Development*, Balasubramanian, R., Santos, V.J. (eds). Springer, Springer International Publishing Switzerland, Cham.

Loomis, J.M. and Lederman, S.J. (1986). Tactual perception. In *Handbook of Human Perception and Performance*, Boff, K.R., Kaufman, L., Thomas, J.P. (eds). John Wiley & Sons, Hoboken, NJ.

Kammerl, J. (2012). Perceptual haptic data communication for telepresence and teleaction. PhD Thesis, Technische Universität München, Munich.

Kane, B.J., Cutkosky, M.R., Kovacs, G.T.A. (2000). A traction stress sensor array for use in high-resolution robotic tactile imaging. *Journal of Microelectromechanical Systems*, 9(4), 425–434.

Kappassov, Z., Ramon, J.A.C., Perdereau, V. (2015). Tactile sensing in dexterous robot hands – Review. *Robotics and Autonomous Systems*, hal-01680649, 74, 195–220.

Kokkonis, G., Psannis, K.E., Roumeliotis, M., Kontogiannis, S., Ishibashi, Y. (2012). Evaluating transport and application layer protocols for haptic applications. *2012 IEEE International Workshop on Haptic Audio Visual Environments and Games (HAVE 2012) Proceedings*, pp. 66–71.

Kokkonis, G., Psannis, K.E., Kontogiannis, S., Nicopolitidis, P., Roumeliotis, M., Ishibashi, Y. (2018). Interconnecting haptic interfaces with high update rates through the internet. *Applied System Innovation*, 1(4) [Online]. Available at: https://www.mdpi.com/2571-5577/1/4/51.

Lee, M.H. and Nicholls, H. (1999). Tactile sensing for mechatronics a state of the art survey. *Mechatronics*, (9), 1–21.

Lee, J. and Payandeh, S. (2012). Signal processing techniques for haptic data compression in teleoperation systems. *2012 IEEE Haptics Symposium (HAPTICS)*, pp. 371–376.

Lowe, D.G. (1999). Object recognition from local scale-invariant features. *Proceedings of the Seventh IEEE International Conference on Computer Vision*, 2, pp. 1150–1157.

Lucia, S., Paolo, G., Watt, S.J., Valyear, K.F., Zuher, F., Fulvio, M. (2019). Active haptic perception in robots: A review. *Frontiers in Neurorobotics*, 13, 53 [Online]. Available at: https://www.frontiersin.org/article/10.3389/fnbot.2019.00053.

Luo, S., Bimbo, J., Dahiya, R., Liu, H. (2017). Robotic tactile perception of object properties: A review. *Mechatronics*, 48, 54–67.

Malley, M.K.O. and Gupta, A. (2008). Haptic interfaces. In *HCI Beyond the GUI: Design for Haptic, Speech, Olfactory, and Other Nontraditional Interfaces*, Kortum, P.. Morgan Kaufmann, Amsterdam.

Mauve, M., Hilt, V., Kuhmünch, C., Vogel, J., Geyer, W., Eefflsberg, W. (2000). RTP/I: An application level real-time protocol for distributed interactive media [Online]. Available at: https://wwwcn.cs.uni-duesseldorf.de/publications/publications/library/Mauve2001a.pdf.

Mauve, M., Hilt, V., Kuhmunch, C., Effelsberg, W. (2001). RTP/I-toward a common application level protocol for distributed interactive media. *IEEE Transactions on Multimedia*, 3(1), 152–161.

Millman, P.A. and Colgate, J.E. (1991). Design of a four degree-of-freedom force-reflecting manipulandum with a specified force/torque workspace. *Proceedings 1991 IEEE International Conference on Robotics and Automation*, 2, 1488–1493.

Minsky, M.D.R. (1995). Computational haptics: The sandpaper system for synthesizing texture for a force-feedback display. PhD Thesis, MIT Press, Cambridge, MA.

Mizuochi, M. and Ohnishi, K. (2013). Optimization of transmission data in bilateral teleoperation. *The Transactions of the Institute of Electrical Engineers of Japan*, 133(3), 314–319.

Nakano, T., Uozumi, S., Johansson, R., Ohnishi, K. (2015). A quantization method for haptic data lossy compression. *2015 IEEE International Conference on Mechatronics (ICM)*, pp. 126–131.

Nasir, Q. and Khalil, E. (2012). Perception based adaptive haptic communication protocol (PAHCP). *2012 International Conference on Computer Systems and Industrial Informatics*, pp. 1–6.

Nitsch, V., Kammerl, J., Faerber, B., Steinbach, E. (2010). On the impact of haptic data reduction and feedback modality on quality and task performance in a telepresence and teleaction system. *International Conference on Human Haptic Sensing and Touch Enabled Computer Applications*. Springer, Berlin, Heidelberg.

Ohka, M. (2007). Optical three-axis tactile sensor. *Mobile Robots: Perception and Navigation*. IntechOpen, London.

Osman, H.A., Eid, M., Saddik, A.E. (2008). Evaluating alphan: A communication protocol for haptic interaction. *2008 Symposium on Haptic Interfaces for Virtual Environment and Teleoperator Systems*, pp. 361–366.

Otanez, P.G., Moyne, J.R., Tilbury, D.M. (2002). Using deadbands to reduce communication in networked control systems. *Proceedings of the 2002 American Control Conference (IEEE Cat. No.CH37301)*, 4, 3015–3020.

Parkes, A., Poupyrev, I., Ishii, H. (2008). Designing kinetic interactions for organic user interfaces. *Communications of the ACM*, 51(6), 58–65.

Phung, M.D., Tran, T.H., Van Thi Nguyen, T., Tran, Q.V. (2012). Control of internet-based robot systems using multi transport protocols. *2012 International Conference on Control, Automation and Information Sciences (ICCAIS)*, pp. 294–299.

Ping, L., Wenjuan, L., Zengqi, S. (2005). Transport layer protocol reconfiguration for network-based robot control system. *Proceedings. 2005 IEEE Networking, Sensing and Control*, pp. 1049–1053.

Robertson, B.E. and Walkden, A.J. (1985). Tactile sensor system for robotics. *Measurement and Control*, 18(7), 262–265 [Online]. Available at: https://doi.org/10.1177/002029408501800703.

Rossi, D.D. (1991). Artificial tactile sensing and haptic perception. *Measurement Science and Technology*, 2(11), 1003–1016 [Online]. Available at: https://doi.org/10.1088/0957-0233/2/11/001.

Schiele, A. (2014). An exoskeleton to remote-control a robot tedxrheinmain [Online]. Available at: https://www.youtube.com/watch?v=JvAho9tym4A.

Schmidt, R.F. (ed.) (1986). Somatovisceral sensibility. *Fundamentals of Sensory Physiology*, Springer, Berlin, Heidelberg.

Schulzrinne, H., Casner, S., Frederick, R., Jacobson, V. (1996). RTP: A transport protocol for real-time applications. Report, RFC 1889.

Shahabi, C., Kolahdouzan, M.R., Barish, G., Zimmermann, R., Yao, D., Fu, K., Zhang, L. (2001). Alternative techniques for the efficient acquisition of haptic data. *SIGMETRICS Performance Evaluation Review*, 29(1), 334–335.

Shahabi, C., Ortega, A., Kolahdouzan, M.R. (2002). A comparison of different haptic compression techniques. *Proceedings of the IEEE International Conference on Multimedia and Expo*, 1, pp. 657–660.

Shirmohammadi, S. and Georganas, N.D. (2001). An end-to-end communication architecture for collaborative virtual environments. *Computer Networks*, 35(2), 351–367 [Online]. Available at: http://www.sciencedirect.com/science/article/pii/S1389128600001869.

Stewart, R. (2007). Stream Control Transmission Protocol. Proposed standard, RFC 4960, September.

Sutherland, I.E. (1965). The ultimate display. *Proceedings of the IFIP Congress*, pp. 506–508.

Torres-Jara, E., Vasilescu, I., Coral, R. (2006). A soft touch: Compliant tactile sensors for sensitive manipulation. Technical Report MIT-CSAIL-TR-2006-014, Massachusetts Institute of Technology's CSAIL, Cambridge, MA.

Uchimura, Y. and Yakoh, T. (2004). Bilateral robot system on the real-time network structure. *IEEE Transactions on Industrial Electronics*, 51(5), 940–946.

Wirz, R., Ferre, M., Marín, R., Barrio, J., Claver, J., Ortego, J. (2008). Efficient transport protocol for networked haptics applications. *Haptics: Perception, Devices and Scenarios, Lecture Notes in Computer Science*, vol. 5024, Springer, Berlin, Heidelberg.

Wirz, R., Marin, R., Ferre, M., Barrio, J., Claver, J.M., Ortego, J. (2009). Bidirectional transport protocol for teleoperated robots. *IEEE Transactions on Industrial Electronics*, 56(9), 3772–3781.

Xie, Y., Chen, C., Wu, D., Xi, W., Liu, H. (2019). Human-touch-inspired material recognition for robotic tactile sensing. *Applied Sciences*, 12, 2537.

Zhang, H. and So, E. (2002). Hybrid resistive tactile sensing. *IEEE Transactions on Systems, Man, and Cybernetics, Part B (Cybernetics)*, 32(1), 57–65.

8

Mapping Wireless Networked Robotics into Tactile Internet

Nicola Roberto ZEMA and Tara ALI-YAHIYA

Department of Computer Science, University of Paris-Saclay, France

A set of robots whose functionalities as a whole and as single elements are enhanced and enabled by wireless networking is called a Robotic Network and enclosed in the framework of Wireless Networked Robots (WNR). In this kind of network, the nodes need to exchange a large set of different types of data, ranging from positional information to large chunks of sensor data and even high quality video. The data also has to be delivered inside the temporal limits associated with the robotic control algorithms.

As the Tactile Internet (TI) is one of the few standards that considers the delivery of kinesthetic and large-bandwidth data among distant entities and inside strict temporal limitations, it is possible to map Wireless Networked Robots inside the Tactile Internet Architecture and Interfaces. This chapter deals with the aforementioned integration by first characterizing Wireless Networked Robots, describing the traffic generated by this kind of network and then translating the requisites, scenarios and uses cases into the Tactile Internet.

8.1. Wireless networked robots

It is possible to define Networked Robotics as a *system of systems* that has autonomous capabilities and network-based cooperation. The autonomous capability refers to the possibility for a system element to move and interact within and with its surrounding physical environment, and the network-based cooperation consists of the global system's extreme dependency on communications, within its composing elements, to operate.

A set of robots displaying these characteristics form a Robotic Network, a subset of Sensor and Actuator Networks (SANs). With respect to SANs, RNs have the added features of degrees of mobility and are generally described as an ensemble, a fleet or a swarm. The need to be mobile and the constraints therein almost automatically mean Wireless Networking is almost the only possible choice for Networked Robotics, thus transforming them to Wireless Networked Robots. The definition has come a long way and it can be traced back to the IEEE Technical Committee (Isler *et al.* 2015) that was originally created for Internet-teleoperated robots and came to include Robotic Networks and Cloud Robotics. Even though the standard did not define them yet, usually the general architecture of WNRs includes a set of entities external to the robot fleet: a Remote Station. The Remote Station is the pivot point between the robots and the external world. The robots send their data to this station and can receive tasks, command and situational updates from it. The communication between the set of robots and the station is usually delegated to a set of gateway robots. These robots collect and manage the traffic from their peers and can relay it to the Remote Station.

An interesting example of WNR is a coordinated fleet instructed to film a sporting event (Natalizio *et al.* 2019). Starting from the case where a single UAV is being remotely operated to film the sport event, the scenario is expanded with first a fleet of independent UAVs, then with a UAV fleet whose elements share their pose and status, and then with the capability to relay the videos to each other, as shown in Figure 8.1. Finally, the following features are added. Each UAV is capable of:

– identifying the areas of interest to be filmed automatically;

– placing itself in order to always be ready to reach the *closest* areas of interest;

– placing itself in order to maximize the quality of the video being relayed to a remote element.

From the case where a single UAV is remotely operated, the importance of networking is increased manyfold as the communications between the robots becomes integral to the system functioning and the network parameters are involved in the algorithms that permit the advanced capabilities.

By considering WNRs as a whole there is a slight paradigm shift in the research focus. From considering the single-robot as the highest hierarchical element of the possible frameworks, now it has become one of the component elements of the ensemble, fleet or swarm. Historically, the research focus has been on how to introduce coordination and cooperation, among the component robotic elements, in terms of movement algorithms, task allocation problems and formation coordination, just to cite a few. However, the approach has almost always been to consider the networking part of WNRs as the set of constraints to adhere to or the problem to overcome. Even with the advent of Controlled Mobility (Kansal *et al.* 2004; Natalizio and Loscrí 2013; Zema *et al.* 2016), there is still no strict definition of what the expected traffic within WNRs is, how to classify and how to treat it.

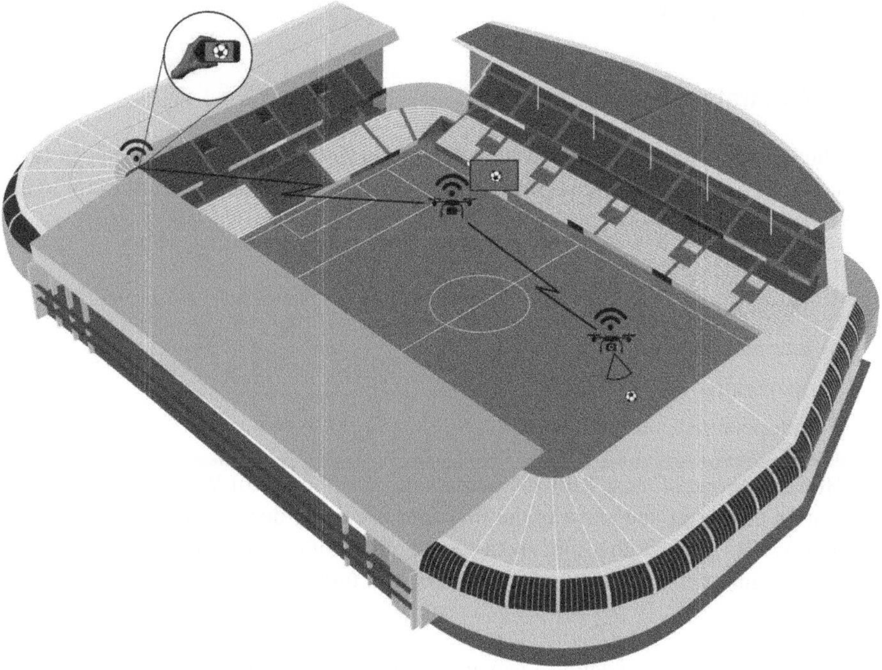

Figure 8.1. *A subset of two robots in a WNR that is used to relay the video of a sporting event to a spectator too far away to receive it in HD from the* filmer *UAV using a second one, a* relayer. *The* filmer *UAV is capable of tracking the sporting event and the* relayer *UAV is capable of placing itself in a position to relay the video, in HD, to all the spectators on its side. For a color version of this figure, see www.iste.co.uk/ali-yahiya/tactile.zip*

In this chapter, we propose to invert the trend, define the WNR traffic features in strict terms and then place them within the framework of the Tactile Internet. This, in turn, will give researchers and implementers the possibility to use WNRs and their networking as a dimension in their works rather than a hard limitation.

The hard constraints of TI traffic classes and Key Performance Indicators specifically constructed for kinesthetic control and remote operation are excellent candidates to be transposed to the world of WNRs. The short delays and low error rates necessary in the Tactile Internet are similar to the ones needed for the coordination of distant robots moving autonomously.

This chapter is subdivided as follows. Section 8.2 describes a new use case in TI for WNRs and describes the kinds of network traffic in WNRs. Section 8.3 focuses on defining the KPIs of the main traffic type of WNRs. Section 8.4 uses the WNR KPIs to map the WNR elements into the TI architecture and interfaces and, finally, section 8.5 concludes the chapter.

8.2. WNR traffic requisites

The Tactile Internet is defined by a standard (Holland *et al.* 2019) that, in turn, defines a set of use cases and related networking constraints. The most relevant to the WNR world are as follows:

1) teleoperation;

2) automotive;

3) Internet of Drones;

4) cooperative automated driving;

5) interpersonal communications.

However, none of these fully encompass the specificities of WNR. First, the strict separation of master and slave in the first four use cases does not adapt well to each robot in WNRs being both a sink and a source of data at the same time, summarized in use case 5 (interpersonal communication), that in turn does not have the specificities of other cases.

It is then necessary to define a new use case specific for WNRs by first taking into account the specific traffic generated and consumed by them.

8.2.1. *Types of traffic in WNRs*

WNRs need specific care as, intrinsically, they include two broad categories of traffic:

– data traffic;

– control traffic.

Data traffic can be broadly characterized as the data necessary or produced for and by the application or task the robots are running. An example is the audio feed collected by an underwater robot's sonar that is transmitted back to a remote station or the Internet traffic of a mobile cellular repeater mounted on a UAV. Control traffic on the other hand is the traffic necessary to the inner working of the WNR, like the telemetry data for reciprocal distance estimation in a robotic fleet or the flying control commands for the leader in a UAV swarm. Given their features, it is necessary to treat them differently and separately as they do have different specificities.

A significant example that can be used to illustrate these different specificities can be found in a *Mobile Cellular Infrastructure* scenario (Boss *et al.* 2016; Jalali 2018; Fotouhi *et al.* 2019). In this scenario, a fleet of UAVs carrying cellular equipment is dispatched and dispersed in an area with the task to cover it. As shown in Figure 8.2, by *covering* we mean that the robots have to displace themselves in order to, at the same time, (i) cover the entire area with the combination of their cellular transceiver communication ranges and (ii) remain in relative contact in order to relay the user data on the ground to each other and then to a *gateway* element, connected to the backbone, by moving inside limited areas according to a set of algorithms.

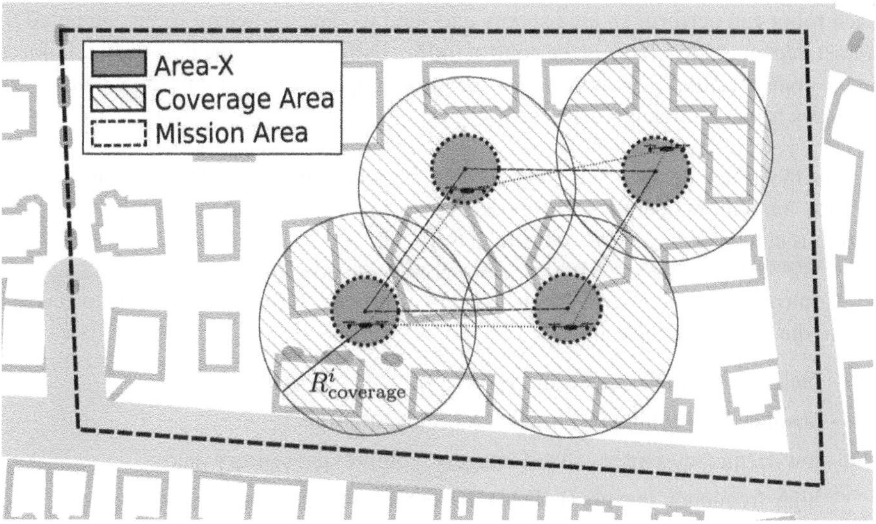

Figure 8.2. *Example of a WNR for* Mobile Cellular Infrastructure *inside its Mission Area. The single UAV with its $R^i_{coverage}$ tries to maximize the fleet Coverage Area by moving inside Area-X while still keeping the formation. For a color version of this figure, see www.iste.co.uk/ali-yahiya/tactile.zip*

8.2.1.1. *Data*

Data traffic refers to the traffic generated by or by virtue of the application running on the networked robots. When the fleet is given a task alongside an application, the data related to the task execution is the data traffic. In the example scenario, the user traffic coming from the backbone to the users on the ground, and vice versa, is the *data traffic*. Although important as a payload, it has almost no direct effect on the fleet operation and it can be physically separated. However important and subject to different shaping and restrictions (phone calls vs. best effort Internet traffic), failures in data traffic delivery do not jeopardize fleet and robot operations. In general, it is not possible *a priori* to characterize this traffic as it is application dependent. It is, however, possible to provision it using the application specifications.

8.2.1.2. *Control*

Control traffic refers instead to the traffic generated by and from the actual fleet and its component elements. The message that concerns task assignment to a specific robot, its feedback, formation coordination and robot remote operation makes up the control traffic.

In the example scenario, when the UAVs enter the area, they are dispatched to specific locations. This dispatch can come from a centralized entity or even by the fleet self-organizing after a local survey. In whichever case, the robots need to communicate with each other or with a distant entity in order to know where to go. In some cases, each robot can perform an area survey and send its data elsewhere to be computed to adjust fleet operation. This is the control traffic. Different from the data traffic, failures in the control traffic do jeopardize fleet and robot operations and thus the whole traffic category has to be treated differently.

As the robot operation is supposed to be autonomous, in the specific case of WNRs, the remote operation of robots is considered to be handled at application level and thus can be deemed part of the data traffic.

Control traffic can instead be enclosed in more strict terms. It is possible to classify it into the following sub-categories:

- inertial kinesthetic (pose and velocity);
- absolute positioning (GPS, GLONASS, Galileo);
- low-frequency sensors (fixed imagery, compressed sensory data);
- high-frequency sensors (LIDAR, RADAR, video imaging, audio capture).

8.3. Traffic shaping and TI haptic codecs

To treat traffic differently, it is necessary to define a way to identify it and then to apply discriminating actions to the identified parts. Given that in general the WNR traffic will lie on an IP basis, it is likely that the already established mechanisms for IP traffic shaping will be put in place, and these are summarized in the Service-Level Agreements of the Quality of Service paradigm applied to the component networks of the whole WNR. However, there is no SLA definition for WNRs as the work of the related IEEE Technical Committee has not progressed further. In this chapter, we thus propose to define the WNR traffic in terms of TI infrastructure and codes, in order to map the necessities of WNRs in an already existing established framework. TI has been chosen as it is the only one that combines either low-latency, strictly high-latency traffic and high bandwidth in a single set of definitions.

8.3.1. *Introduction*

Different, however, from the standard TI use cases, for WNRs there is an added challenge: the presence, at the same time, of prioritized traffic for general operation and prioritized payload (the data traffic).

Referencing the mobile cellular station scenario previously introduced in section 8.2.1, hierarchies can be defined inside control traffic, where inter-robot telemetry traffic has priority over remote operation traffic, going out to the backbone to a remote server, that in turn has priority over low-rate traffic that is used, for instance, in the algorithms for fleet maintenance. Inside the data traffic, more *classic* hierarchies can be maintained, with tolled phone calls having precedence over best-effort Internet.

Lots of questions arise at this point such as if it is necessary to include all the hierarchies together or if it is better to separate them and leverage physical layer separation where available.

Different strategies could be put into operation. One of them is to treat control traffic as always having higher priority over data traffic. Another is to carefully optimize interleaving traffic and control on a per-scenario basis. A third solution would be to define novel algorithms and solutions that are capable of dynamically changing priorities according to the environment. In any case, this research line is outside the scope of this work, that is the preliminary step of defining these traffic classes and their sub-classes.

8.3.2. *Mapping WNR control traffic to TI*

The first step of this analysis is to clearly define the WNR control traffic. As previously defined in section 8.2.1.2, control traffic can be subdivided into sub-categories that collect together functionalities with the same requirements.

8.3.2.1. *Kinesthetic/inertial and absolute positioning*

The first and maybe most important traffic class is the one that carries inertial/kinesthetic information. The mobility is the most important degree of freedom in WNRs and thus follows that information regarding this is to be treated with the highest reliability and speed. In the cases where the piloting is done remotely either with a human-in-the-loop (remote piloting) or machine-in-the-loop (remote formation control), it is necessary that updates in robot position and velocity are relayed to a given entity with the minimum possible tolerance. Failure to do so could well result in equipment damage and divergence in maintenance algorithms. Thus, in these cases, the traffic carrying this information would need the highest priority class. Even when the piloting is completely autonomous, it is still necessary to share, among the robots, the information about pose and velocity to run, for instance, keeping formation. This case is also exemplary for another requirement: the link bidirectionality. Not only does a robot send data outside, but it is expected to receive navigational commands or even direct inputs to its control loop. In any case, the highest environmental dynamicity is expected. According to Matheus and Königseder (2017), the necessary bandwidth is narrow and is dependent of the degrees of freedom involved. The subclass is called inertial as the data is generated, for each robot, by its own instruments and internal navigation systems (inertial measurement unit, altimeter) and it is differentiated by the data generated by absolute positioning systems (GPS, GLONASS and Galileo). This is due to the update frequencies of absolute positioning being different, and lower, from the inertial ones. Furthermore, absolute positioning could not always be present and active all the time. In the following table, it is possible to sample the traffic features.

Type	Latency	Burst size	Reliability	Data rate
Kinesthetic/inertial	1–10 ms	2–8B ∀ DoF	99.999	1–4 k pkt/s
GPS, GLONASS, etc.	30–40 ms	2 kB	99.9	5–512 kbps

8.3.2.2. *Low-frequency sensors*

With the WNR algorithms becoming increasingly complex, they often need more information than kinesthetic data. An underwater robot exploring a cave could use a SONAR system to scan its surroundings. If the robot is located at the end of a *chain* of other robots that relay communications to a distant station, it could also send the sonar imagery to the station. In respect to the previous sub-category, the traffic that carries the imagery has a greater bandwidth and also a slightly lower reliability limit. The sonar images are generated periodically, they do not constitute a constant stream, and they can be compressed and passed through an error correction algorithm. A whole broad category of information falls within these boundaries and includes, for instance, samples of connected users (and their statistics) in the mobile cellular station scenario or samples of detected chemical data when UAVs are used to survey pollutants. Robots can also be consumers of this kind of traffic. WNR algorithms could include fog or even local-based computational distribution, and thus, the traffic is always bidirectional: in the underwater robot scenario, the robots could share the SONAR information and cooperate computationally to apply filters and analysis to the imagery. This traffic can be assimilated to the one often common in vehicular networks and, specifically, in scenarios of low or absent dynamicity. In this case, the following table can be used as a reference (Cao *et al.* 2016; Campolo *et al.* 2017), where the traffic is further subdivided into low-frequency video, that includes image compression algorithms, and low-frequency aggregate, for uncompressed data.

Type	Latency	Burst size	Reliability	Data rate
Low-frequency aggregate	10–50 ms	~2.4 kB	99.9	10–40 Mbps
Low-frequency video	50–150 ms	~2.4 kB	99.9	1–10 Mbps

8.3.2.3. *High-frequency sensors*

WNRs can also rely on direct computation of video data to run the algorithms (Natalizio *et al.* 2019), and there are cases where high-bandwidth data analysis is put inside the robot control loop. For instance, in a distributed UAV surveillance scenario, the robots are dispersed over an area and capture video that is then distributed to the WNR component for cooperative evaluation. The video quality, in terms of perceptive metrics like PSNR, is used to tune target network parameters like jitter and latency between the video source and its sink. As the scenario is a wireless one, these parameters are then translated into a set of constraints on the robot's relative positions. It then forms a closed control loop that involves robot positioning and video quality. The same reasoning could be applied to a high-frequency sonar feed in an underwater exploration scenario. High-bandwidth flows have different features with respect to the other cases, that translate into even more stringent limitations when it is necessary to maintain bidirectionality of links. In a distributed surveillance scenario robots could cooperate in the video analysis to detect intrusion without relying on external computation resources and the WNR should be capable of sharing the video feed among robots (Muzaffar *et al.* 2020; Marshall *et al.* 2008). Based on these features, the following table can be composed, where for flying robots even more stringent boundaries are set, to take into account the increased environmental dynamicity.

Type	Latency	Burst size	Reliability	Data rate
UAVs	30–40 ms	4 kB	99.999	1–100 Mbps
Rovers	10–100 ms	~2.4 kB	99.9	1–10 Mbps

8.3.2.4. *Tactile Internet codec translation*

The TI task subgroup group IEEE 1918.1.1 (Steinbach *et al.* 2018) is bound to define the codecs for the proposals. In this case, the codecs correspond to the definitions of network-side support for the TI. Two main codec categories are identified: tactile and kinesthetic and the stated objective is to reduce the average packet rate in haptic communications in order to make the two endpoints of a Tactile Internet communication appear to be locally available even in the presence of delays.

The tactile codecs correspond to the encoding and transfer of the information related to human perception of touch while the kinesthetic codecs are related to the information about limb movement in humans.

The kinesthetic codecs are of great importance for WNRs as the kinesthetics referred to overlaps with the kinesthetic/inertial information referred to in section 8.3.2.1. Even though the standard does not yet fully define the codecs, an important groundwork has been laid in identifying classes of traffic and the way they should be treated. The standard divides the codecs into *delay-tolerant* and *delay-intolerant* ones. This difference stems from the fact that it is necessary to put in place a stabilizing control mechanism when in the presence of significant delays, usually more than 5 ms. When the expected or detected latency is below the threshold, then the *delay-intolerant* codecs are run, to compress the transmitted information and reduce the number of exchanged packets in the network. When the latency is above the threshold, then the objective is to stabilize the system. The investigated alternatives so far include:

– algorithms that exploit the human perception of kinesthetic stimuli (Weber's law (Culbertson *et al.* 2014));

– algorithms based on data reduction.

In WNRs, Weber's law algorithms need to be carefully analyzed before being adopted as they are designed for anything other than machine-in-the-loop systems and the data reduction algorithms show performance degradation when used in highly dynamic environments (Steinbach *et al.* 2018).

For these difficulties, to date, kinesthetic codecs have been explored but not yet defined in the standard and delay compensation mechanisms are of great interest for WNRs.

More important than delay is the loss compensation. In the wireless networks provisioned, loss probabilities are non-negligible and mechanisms need to be put in place to compensate for message losses. An example of compensation mechanism (in this case, for the packet loss) is described in Manfredi *et al.* (2020). In this case, a layer of redundancy is added in the packet flow that handles the formation control of an UAV fleet to take into account possible burst losses.

8.4. WNRs in the Tactile Internet architecture

According to the previous sections is it now possible to define a new TI use case for WNRs and populate it with a set of KPIs. One of the most striking features with respect to the other use cases is that in WNRs the traffic and the TI encasing it is bidirectional. It has to be provisioned that each robot could communicate with any of its peers via a TI set of interfaces. Table 8.1 summarizes the use case.

Type	Latency	Burst size	Reliability	Data rate
Kinesthetic/inertial	1–10 ms	2–8B ∀ DoF	99.999	1–4 k pkt/s
Absolute positioning	30–40 ms	2 kB	99.9	5–512 kbps
Low-frequency aggregate	10–50 ms	~2.4 kB	99.9	10–40 Mbps
Low-frequency video	50–150 ms	~2.4 kB	99.9	1–10 Mbps
UAVs HD video	10–100 ms	~2.4 kB	99.9	1–100 Mbps
Robot high frequency	30–40 ms	4 kB	99.999	1–10 Mbps

Table 8.1. *WNR use case for TI features*

The KPIs can then be used to map WNRs into the TI framework interface features and architecture.

8.4.1. *WNRs in the TI architecture and interfaces*

Tactile Internet architecture defines a set of entities and interfaces that are connected together and evaluated with a set of KPIs described by the use case the TI is deploying at that moment. After having translated the WNR traffic requirement into a set of KPIs, it is now possible to use them to map the WNR into the TI architecture and interfaces, translating each part of the WNR into the most appropriate part of TI. To better present the mapping, it is possible to leverage again the *Mobile Cellular Infrastructure* scenario of section 8.2.1 and shown again in Figure 8.3. In this scenario, a set of UAVs, a swarm, is dispatched to distribute itself in a given area where mobile users are located. The UAVs carry three network interfaces:

– a portable cellular base station that has its own range;

– a network interface for the UAVs to relay the mobile user data;

– a network interface for the WNR operation.

It also deploys an *intermediate station*, i.e. the UAV carrier, a local base station, that is in communication, at the same time, with the UAVs, or a subset of them, and with a *remote base station* using a long-range link.

For the moment, we focus on the WNR operation and suppose that the user data, the data traffic of section 8.2.1.1, is outside the scope of this chapter. While not using the same frequencies and technologies, the UAVs could leverage the presence of the *intermediate station* to reach the *remote base station* for the *data traffic*. In any case, it is necessary to at least provision the interfaces for just the control traffic and then add on top on them the per-application requirements of the data traffic.

Among the UAVs, a subset of them is selected to act as relay points with the base station, to be used by the ones too far away to be in direct communication, or perhaps without a sufficient link quality, with the *intermediate station*. These *gateways*, other than relaying the selected traffic, collect and aggregate information to be sent to the *intermediate station*.

The WNR is running an algorithm that automatically distributes the UAVs according to a set of parameters, which could be the link quality among specific couples, the load from the users on the ground and the resources of the specific UAV. The single algorithm, or a set of them, could be either centralized, distributed or even isolated, and thus, the *decisions* could be taken at the swarm, *intermediate station* or even *remote base station* level. Load balancing decisions could be taken remotely, using optimization algorithms that need large computational resources and formation keeping can be done directly inside the swarm.

In this scenario, the UAVs continuously exchange their movement vectors and samples of their estimated absolute position among themselves, as well as aggregate values for their low-frequency sensors (estimates of the connected users, load and resource information, battery power left). Aggregates of this data could be sent from the *gateway* UAVs to the *intermediate station*. Selected UAVs could also activate their high-frequency sensors (i.e. cameras) and send the data to the *intermediate station* or to the *remote station*.

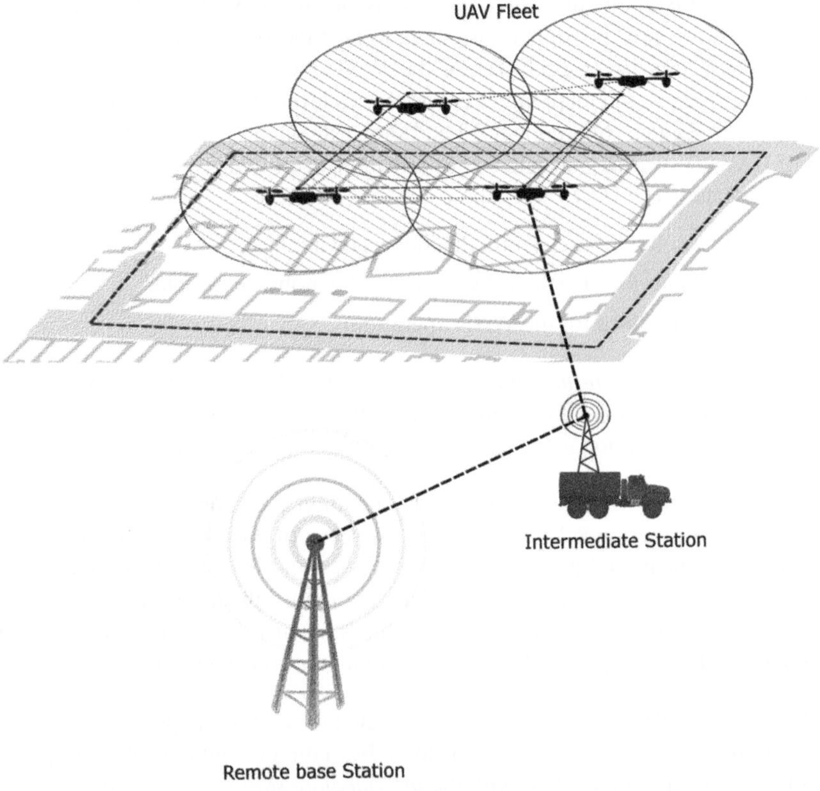

Figure 8.3. *A WNR implementing the* Mobile Cellular Infrastructure *scenario. The UAVs are distributed in the area and are in contact with each other and the* intermediate station *provides a link with a* remote base station

8.4.1.1. *Functional architecture*

The basic TI entity, the tactile device, that contains sensors, actuators and their *nodes*, can be immediately translated to a mobile robot. A robot can contain, in fact, its navigation sensors and the engines are actuators. The fact that there is usually a central processing unit connected to them via a bus makes the integration of TD, sensors and actuator nodes seamless in a robotic entity inside the TI.

An initial mapping is almost straightforward. In WNRs, a tactile edge, composed by the various TDs, is the robot swarm, that has a remote control station counterpart in another tactile edge, remotely located, as shown in Figure 8.4. As the *remote base station* is supposed to have full control over the UAVs, it is identified as another tactile edge with its own controllers (not shown in the figure). The placement of the gateway node and the network controller, or together in the gateway-node controller, needs extra care in WNRs due to the peculiar nature of WNRs. The GNC of a WNR can be located in each robot as part of the TD and also in the *intermediate station*, according to the possibilities described in Holland *et al.* (2019). This vehicle's capabilities vary wildly according to the application, and thus, it is important to decide if the intermediate station belongs to the tactile edge or not and act accordingly per-application. The same reasoning applies to the tactile service manager, that can be placed in the swarm or remotely.

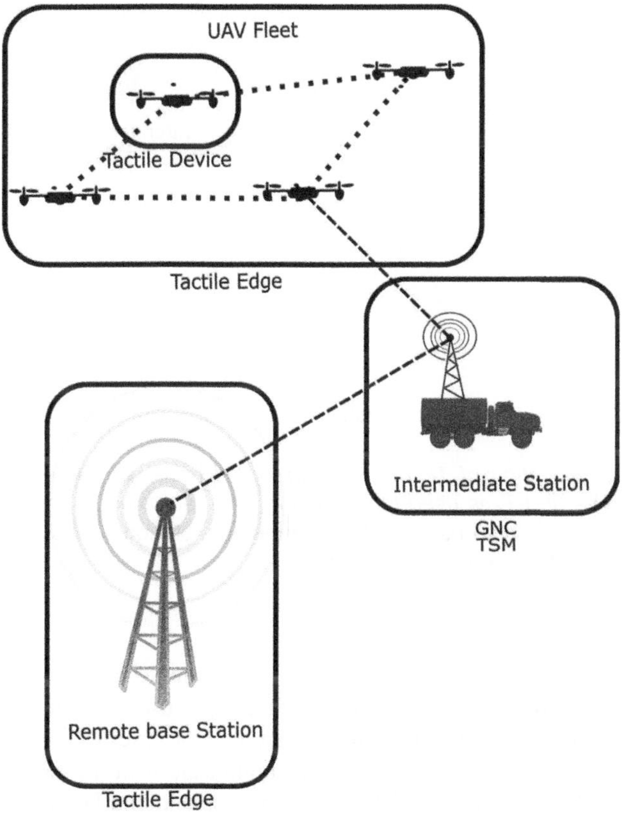

Figure 8.4. *Identification of the TI infrastructure in a WNR scenario. The* remote base station *and the UAVs themselves represent two tactile edges and the GMC and TSM is located inside the* intermediate station

The placement and mapping of the support engine is also complex. In WNRs, the selection of a wide set of algorithms, that range from network load balancing to task assignment, can be done distributedly. In this case, differently from the TI standard, each TD contains the SE or a part of it. In the other cases, the SE can be located in the remote station or in the intermediate station.

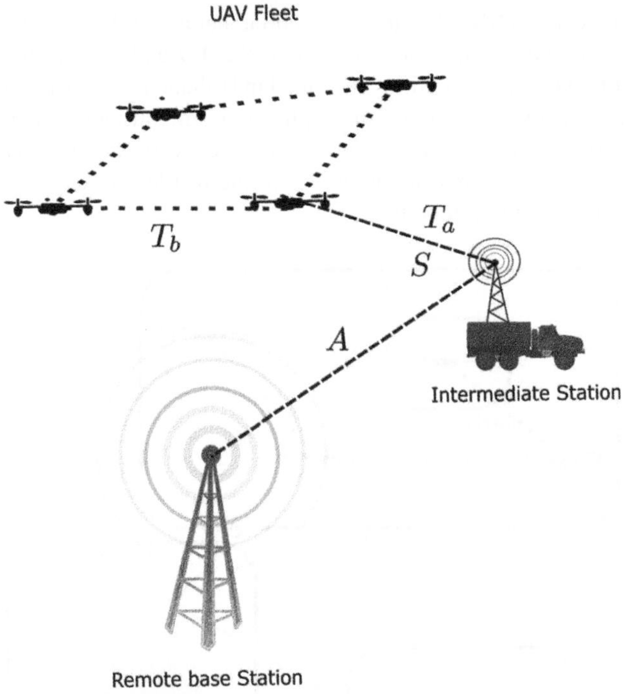

Figure 8.5. *Identification of TI Interfaces in a WNR scenario. Interface T_b is used among the UAVs and T_a is used to connect it, as a tactile edge, to the* intermediate station. *Interface S is also used as the* intermediate station *also contains the TSM. Interface A connects the* remote base station.

8.4.1.2. *Interfaces*

If all the robots inside a WNR are TDs, to communicate among each other the interface to be used is the T_a, while interface A is used by selected nodes (the gateways) to communicate with the remote station. In the cases where the intermediate station is deployed, then the robots could communicate with it using a T_b interface. This is different from the standard TI as it is defined that all the TD have an A interface active. The identification of the interfaces S and O is complex. As the SE can be distributed among nodes, being placed in the intermediate station or even in the remote station, the interfaces S and O could be merged with the T_a or even T_b or A.

8.4.1.3. *Bootstrapping*

The literature (Holland *et al.* 2019) defines a set of protocols to bootstrap the system. These protocols have been designed with the aim of providing different solutions to TI bootstrapping in case of different scenarios. The protocols describe the steps for the device and interfaces to follow in order to activate the Tactile Internet in a scenario. Some of them are unsuitable for WNR operations as these include phases where the system is in a dormant state (the robots are inside the *intermediate station*, moving towards the area of interest of the example scenario) and long-range wireless links that could be subject to disruption. Furthermore, the WNR is managed, either in a distributed or centralized way, as a whole. The robot operation parameters as defined in advance refer to a machine-to-machine scenario and are often synchronized.

These peculiarities reduce the choice of possible bootstrapping protocols to the *ad hoc TI paradigm*, that assumes no infrastructure is *a priori* online and that the communication is initiated by the tactile edge. In reference to the example, it is supposed that the *intermediate station* arrives in the designated area and then launches the UAVs. While the UAVs fly into their initial position, the tactile devices are activated and connected via the T_b interfaces. Once the tactile edge is formed, it starts the bootstrapping by requesting to communicate with the *intermediate station* and then the *remote station*. The delays and periodic handshakes accounted for in Holland *et al.* (2019) do represent a small resource delta with respect to the necessary control traffic.

8.4.1.4. *KPI mapping*

According to the section content, it is now possible to define the features of the TI interfaces involved according to the needs of the WNR scenario. It is provisioned that T_a interfaces will always have to support the kinesthetic/GPS data among the UAVs, as well as the low-frequency sensor data. Thus, these interfaces will have to support the KPIs of Table 8.1. They will also have to support the low-frequency aggregates. The interface T_a will need to support aggregates of T_b and the extra data the *Gateway* UAVs collect from the swarm, similar to the low-frequency video features of the same table. There would also be events when the high-frequency sensors would be activated on selected UAVs. As per-scenario, it would be necessary to aggregate or create new interfaces in order to transmit increased bandwidth data up to the *intermediate station*. Multiple T_a interfaces from multiple UAVs could be arranged to split and transmit the high-frequency data and multiple *gateways* could connect to the *intermediate station* at the same time. In any case, the A interface should be capable of supporting multiple high-frequency streams as shown in Figure 8.6.

8.5. Conclusion

In this chapter, we have started from the definition of Wireless WNRs, introduced some important examples and then defined the expected traffic inside and through it.

This was done to verify whether it is possible to encase the WNR paradigm into the TI and provide a guideline for researchers and implementers that wish to continue to create and develop wireless networked robotics. We have demonstrated that it is possible to translate WNR traffic features into the KPIs of TI and then use them to map the WNRs inside it. We have identified a new use case for TI to include WNRs, how to map the components of WNRs into the TI framework and what interfaces and bootstrapping protocols to use. We hope that this groundwork could also be used to continue the work on the WNR Technical Committee and its standard.

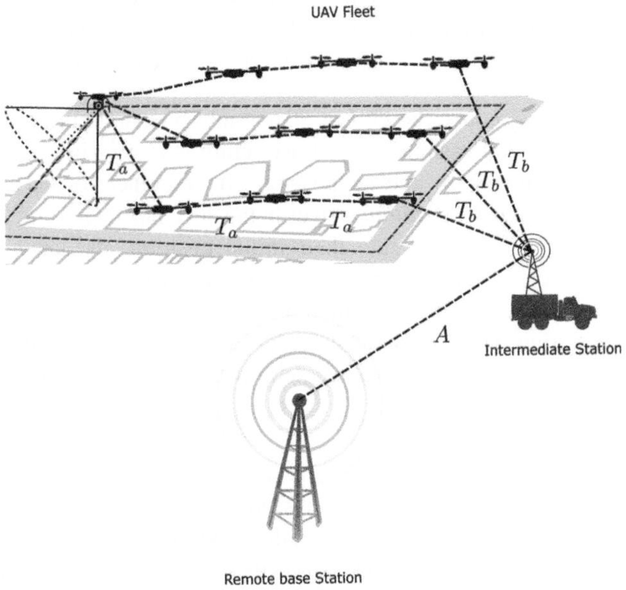

Figure 8.6. *A WNR uses aggregate T_a TI interfaces to support high-frequency sensors*

8.6. References

Boss, G.J., Hamilton, I.R.A., Mukherjee, M., Mukherjee, M. (2016). Deployment criteria for unmanned aerial vehicles to improve cellular phone communications. US Patent 9,363,008.

Campolo, C., Molinaro, A., Iera, A., Menichella, F. (2017). 5G network slicing for vehicle-to-everything services. *IEEE Wireless Communications*, 24(6), 38–45.

Cao, H., Gangakhedkar, S., Ali, A.R., Gharba, M., Eichinger, J. (2016). A 5G V2X testbed for cooperative automated driving. *2016 IEEE Vehicular Networking Conference (VNC)*, IEEE, pp. 1–4.

Culbertson, H., Unwin, J., Kuchenbecker, K.J. (2014). Modeling and rendering realistic textures from unconstrained tool–surface interactions. *IEEE Transactions on Haptics*, 7(3), 381–393.

Fotouhi, A., Qiang, H., Ding, M., Hassan, M., Giordano, L.G., Garcia-Rodriguez, A., Yuan, J. (2019). Survey on UAV cellular communications: Practical aspects, standardization advancements, regulation, and security challenges. *IEEE Communications Surveys & Tutorials*, 21(4), 3417–3442.

Holland, O., Steinbach, E., Prasad, R.V., Liu, Q., Dawy, Z., Aijaz, A., Pappas, N., Chandra, K., Rao, V.S., Oteafy, S., Eid, M., Luden, M., Bhardwaj, A., Liu, X., Sachs, J., Araújo, J. (2019). The IEEE 1918.1 "tactile internet" standards working group and its standards. *Proceedings of the IEEE*, 107(2), 256–279.

Isler, V., Sadler, B., Preuchil, L., Nishio, S. (2015). Networked robots [TC spotlight]. *IEEE Robotics Automation Magazine*, 22(3), 25–29.

Jalali, A. (2018). Broadband access to mobile platforms using drone/UAV background. US Patent 9,859,972.

Kansal, A., Rahimi, M., Estrin, D., Kaiser, W.J., Pottie, G.J., Srivastava, M.B. (2004). Controlled mobility for sustainable wireless sensor networks. *2004 First Annual IEEE Communications Society Conference on Sensor and Ad Hoc Communications and Networks, 2004. IEEE SECON 2004*. IEEE, pp. 1–6.

Manfredi, S., Natalizio, E., Pascariello, C., Zema, N.R. (2020). Stability and convergence of a message-loss-tolerant rendez-vous algorithm for wireless networked robot systems. *IEEE Transactions on Control of Network Systems*, 7(3), 1103–1114.

Marshall, A., Yap, K.M., Yu, W. (2008). Providing QoS for networked peers in distributed haptic virtual environments. *Advances in Multimedia*, Article ID 841590.

Matheus, K. and Königseder, T. (2017). *Automotive Ethernet*. Cambridge University Press, Cambridge, UK.

Muzaffar, R., Yanmaz, E., Raffelsberger, C., Bettstetter, C., Cavallaro, A. (2020). Live multicast video streaming from drones: An experimental study. *Autonomous Robots*, 44(1), 75–91.

Natalizio, E. and Loscrí, V. (2013). Controlled mobility in mobile sensor networks: Advantages, issues and challenges. *Telecommunication Systems*, 52(4), 2411–2418.

Natalizio, E., Zema, N.R., Yanmaz, E., Pugliese, L.D.P., Guerriero, F. (2019). Take the field from your smartphone: Leveraging UAVs for event filming. *IEEE Transactions on Mobile Computing*, 19(8), 1971–1983.

Steinbach, E., Strese, M., Eid, M., Liu, X., Bhardwaj, A., Liu, Q., Al-Ja'afreh, M., Mahmoodi, T., Hassen, R., El Saddik, A., Holland, O. (2018). Haptic codecs for the Tactile Internet. *Proceedings of the IEEE*, 107(2), 447–470.

Zema, N.R., Natalizio, E., Ruggeri, G., Poss, M., Molinaro, A. (2016). Medrone: On the use of a medical drone to heal a sensor network infected by a malicious epidemic. *Ad Hoc Networks*, 50, 115–127.

9

HoIP over 5G for Tactile Internet Teleoperation Application

Tara ALI-YAHIYA[1], Wrya MONNET[2] and Bakhtiar M. AMIN[2]

[1] *Department of Computer Science, University of Paris-Saclay, France*

[2] *Department of Computer Science and Engineering, University of Kurdistan Hewlêr, Erbil, Iraq*

In this chapter, we suggest integrating the IEEE 1918.1 TI architecture into the 5G new radio (NR) access technology and core network, with adequate mapping between the entities and functionalities of both systems. The IEEE 1918.1 working group has undertaken the task of developing a standard for the TI. They have defined the TI architecture along with its associated elements and interfaces. It is worth mentioning that the IEEE 1918.1 standard is considered as an overlay on 5G and other networks. However, this overlay needs some interoperability capabilities that are not defined in the standard and are left open to the operators, depending on the deployed TI use cases (Van Den Berg *et al.* 2017). In this chapter, we propose an overlay architecture design for an integrated TI system, based on the IEEE 1918.1 architecture and protocols with a 5G NR network. In order to support this overlaying, we mapped the functionalities of the IEEE 1918.1 architecture to the functionalities of the 5G core network. Also, we introduced some functionalities that facilitate this interoperability through functional modules that are inserted in the edge and network domains. We propose a conceptual teleoperation system based on 5G NR as access technology to the network. The control of the proposed system requires a low end-to-end delay (E2E) to guarantee a stable system operation and correct haptic feedback. For this purpose, we employed and

The Tactile Internet,

studied the performance of the Haptics over Internet Protocol (HoIP), in order to examine its feasibility to ensure a reliable communication over 5G NR in terms of bit rate and E2E delay. Through numerical analysis and simulations, we show that the HoIP overhead will not violate the quality of service (QoS) requirements in terms of delay.

9.1. Related works

All of the efforts carried out to propose a new architecture for TI did not take the architecture of IEEE 1918.1 into consideration. Instead, each work proposed its own architecture focusing on only one case study. For example, Miao *et al.* (2018) proposed an architecture based on 5G for telesurgery robots with a combination of human–machine interaction data flow and edge computing, as well as the network slicing characteristic in 5G, all with the aim of reducing the interaction time between the robot and cloud computing (Holland *et al.* 2019). Ateya *et al.* (2017a) introduced a 5G network architecture for TI that employs SDN for a QoS guarantee at the core of the network, while enabling NFV. They divided the architecture into several levels, namely physical, software and application and deployed mobile edge computing (MEC) at each level to reduce the E2E delay. In Szabo *et al.* (2015), the SDN was combined with network coding, especially the random linear network coding, to reduce the E2E in the transport domain composed of multi-hop network topology. Ateya *et al.* (2017b) proposed an MEC design for an integrated fiber-wireless (WiFi) access network. They studied different radio access network (RAN) technologies to investigate the performance of the networks in terms of delay, response time efficiency and energy consumption for the devices involved in the edge domain only. Oteafy and Hassanein (2019) claimed that using MEC over TI, with a combination of smart computing and the IoT, would improve the network performance in terms of QoS for different types of applications. They depended on a framework for multi-tiered cognition that takes the contextual information in haptic exchange of information, especially the haptic feedback, into account.

Several research articles suggested 5G architecture as an empowering technology for TI, as in Aijaz *et al.* (2017), Dohler *et al.* (2017) and Antonakoglou *et al.* (2018). These works mainly focused on the design of edges i.e. master and slave domains connected together by a 5G network. They mainly concentrated on the model-meditated teleoperation approach to facilitate the design of a TI framework, while also investigating different protocols for haptic communication in the different layers of the network protocol stack. Maier *et al.* (2018) focused more on the importance of human–machine interaction for automating physical and cognitive human tasks in TI. Li *et al.* (2019) investigated the design of 5G NR for guaranteeing the QoS in terms of URLLC; it mainly dealt with the design of 5G from the perspective of medium-access control and physical layers (MAC/PHY), with particular applications to TI.

Considering only the traditional protocols in the application or transport layers seems to not be a good option for haptic communication. For instance, the user datagram protocols (UDP) and transmission control protocols (TCP) are considered for haptic communication. Such protocols have their own advantages and disadvantages. TCP is considered as a heavy and reliable protocol for haptic communication where peers need to establish a connection in order to communicate. Additionally, TCP introduces high latency, which is not suitable for real-time TI applications. On the other hand, UDP is considered as a light and suitable protocol for haptic communication. It is different from TCP in its reliability, which does not support for haptic (Al Ja'afreh *et al.* 2018; Antonakoglou *et al.* 2018). Some transport protocols were proposed to carry the haptic data, as they were not specifically designed for this purpose. For example, both the synchronous collaboration transport protocol (SCTP) (Shirmohammadi and Georganas 2001) and the light TCP (Dodeller 2004) encapsulated UDP in the protocol stack. However, since UDP is not a reliable protocol, the authors proposed to add the reliable delivery of packets. This is not practical since it added more latency to the communication, and thus is not a realistic solution in the case of haptic communication.

Cen *et al.* (2005) proposed Supermedia Transport for teleoperations over Overlay Networks (STRON), for real-time Internet-based teleoperation systems. More precisely, to control a robot through an operator, through the Internet, a mixture of sensory information remotely controlling robots, video, audio and haptic feedback was used and referred to as a supermedia stream. The performance of the protocol, developed using network simulator-2 (NS2), was compared to TCP and SCTP in terms of delay and packet loss. It was found that it outperformed both protocols for supermedia steams. This is due to the fact that STRON uses multiple disjoint paths and forward error correction encodings to decrease the E2E delay for supermedia streams, thus guaranteeing the QoS for the different types of applications mixed in one stream. The use of the session initiation protocol (SIP) is common in multimedia sessions and conferences with different media such as audio, video and text (King *et al.* 2010). This protocol is widely used to provide many services, while negotiating the QoS parameter for each service at the beginning of the session. King *et al.* (2010) proposed the use of SIP for sending haptic data through the integration of SIP with a haptic codec, which permits the integration of the encoded haptic data within the packet to be transferred through the network. The authors used a robotic arm that emulates a writing hand within LAN. No investigation was conducted on the reliability, packet loss or delay. Another protocol that was compared to TCP and SCTP was the interactive real-time protocol (IRTP). This protocol operates in the application layer, which adds a header length of 28 bytes to each haptic sample that is transmitted (Schulzrinne and Casner 2003). Further, the protocol was used for interactive applications, including audio and video services on the Internet. Moreover, the usage of this protocol was to transport teleoperation data that transmits over the network (Rosenberg *et al.* 2002). The advantage was that it performed better in terms of packet loss and was better in heavy crowded networks with less latency. It was stated that there are still some problems that remain to be solved before it can be applied in haptic data transformation (Rosenberg *et al.* 2002; Schulzrinne and Casner 2003). In addition, the massive overheard that this protocol requires may introduce heavy load in the network (Gokhale *et al.* 2015).

In King *et al.* (2009), a telesurgical robotics project was introduced to test telesurgical data transmission. One of the protocols used to test this process was the interoperable telesurgical protocol (ITP). Such a protocol operates at the application layer to transmit the required haptic data, while using the lightweight UDP as a transport layer. This is due to its light overhead introduced to the network regarding TCP and the traditional UDP. The authors performed a test bed between Japan and United States through the public Internet. The test bed included two master robots (developed independently), which controlled one slave in a remote way. The interoperability of the robot's hardware is investigated through the implementation of ITP. The protocol was proved to work well in a unilateral communication where less degree of freedom (DoF) is deployed in the robot's movement, and only the usual feedback type is used.

Another protocol that was primarily concentrated on haptic communication was called the application layer protocol for haptic networking (ALPHAN) (Osman *et al.* 2007). It operates on the application layer and dynamically adapts itself to the different requirements of QoS. It is implemented upon UDP while adding a reliability concept to it, which is lacking in the original UDP. They used different buffers for different types of applications to give priority to the time-sensitive ones. The context of the implementation was the Internet, and they proved that the protocol performs well in terms of delay and packet loss compared to a single-buffer protocol.

The authors presented a perception-based adaptive haptic communication protocol (PAHCP) above UDP (Nasir and Khalil 2012). It is mainly concerned with haptic exchanges, such as positions, velocities and force values. The protocol was deployed on LAN as the environment of implementation; however, the main focus was on how to reduce the amount of data transmitted over the network through the use of a modified version of just noticeable difference (JND). The main outcome of this work is to test the JND regarding the force, position and velocity, while reducing the number of packets traveling in the network without affecting human perception.

Finally, a protocol known as ADMUX (adaptive multiplexer for haptic–audio–visual data communication) was proposed to perform at the application layer for synchronizing haptic audio-video files on the Internet (Eid *et al.* 2011). Further, the unique feature of this protocol is multiplexing, which allows network resource allocation to diverse media streams. Moreover, experiments have shown that this protocol has provided a dynamic bandwidth allocation based on the network conditions and media types used. In addition, in comparison to previous protocols, this one works better in packet delays of 1 ms for haptic interaction transformation. Furthermore, experiments have shown that this protocol easily adapts and changes with network conditions and application needs. An advantage of this protocol was error detection and correction, especially with media stream packets, packetized elementary streams (PES). Because UDP was used by ADMUX, it was considered as an unreliable communication protocol. Hence, to resolve this issue, it was suggested to use an error flexibility algorithm (Wongwirat *et al.* 2005).

Li *et al.* (2019) proposed an up-to-date PHY/MAC style for NR/LTE URLLC in 5G technologies. In summary, in NR URLLC, the ascendible field (e.g. subcarrier spacing, TTI, round trip time (RTT)) allows NR to own the capability of giving services, with completely different liability and latency needs. In particular, shortening the TTI and hybrid automatic repeated request (HARQ) RTT will considerably improve the system capability under URLLC needs. Moreover, wide-band allocation for URLLC and the dynamic multiplexing of URLLC and alternative traffic are extremely interesting because the URLLC system capability is increasing super-linearly with respect to the accessible information measure. Additionally, the theoretical queuing analysis and system-level simulations have been provided to support these system-style decisions that were being contributed to 3GPP standards. On the other hand, in LTE technology, the URLLC style should follow the same field as gift LTE, thanks to the concept of backward compatibility. Hence, almost like the NR URLLC style, shortening the TTI by reducing the amount of OFDM symbols in one TTI conjointly, improves the system capability. In addition, enhancements on the present LTE management and knowledge channels will facilitate LTE to meet the URLLC requirement. Further, there are many doors that have been left open for future research to develop and find the best security options.

Another paper suggested by Ateya *et al.* (2018) has shown that one of the most stylistic aspects of the TI system is the 1 ms end-to-end latency, which is taken as being the biggest challenge with system realization. Further, as a requirement of the recent developments and capabilities of the 5G cellular system, the TI can become a true method to beat the 1 ms latency to use a centralized controller within the core of the network. Moreover, there is more focus on the core network as the main field of study. Also, this is often the concept behind the software-defined network (SDN). This paper introduces a TI system structure that employs SDN within the core of the cellular network and mobile edge computing (MEC) at multi-levels. Moreover, the work is especially involved with the structure of the core network. Additionally, the system is simulated over reliable surroundings and introduces a spherical trip latency of the order of 1 ms. Therefore, this could be taken by the reduction of intermediate nodes that are involved within the communication method. Mekikis *et al.* (2020) illustrated that industries and firms struggle for smaller business and product increment and there is an in-depth effort from the research community for novel and profitable automation processes. This effort has given rise to the 5G tactile web, which is considered to have very low latency communication in combination with high accessibility, responsibility and security. In this paper, the key technologies to support the characteristics of the Tactile Internet in industrial environments have been examined. Additionally, the implementation of a unique 5G NFV-enabled experimental platform has been illustrated. Therefore, in this paper, it was suggested to use the tactile web for low latency and shorten the business for increasing product efficiently. Also, the capabilities of NFV and SDN needed to satisfy the requirements of this paper have been examined using an NFV-enabled experimental platform that follows the philosophical doctrine framework. The results have shown that sub-millisecond latencies are achievable for services hosted directly at the IoT entry, which are unaffected by network congestion. In addition, it is evidence that the paper has fulfilled the difficult industrial wants. Finally, it is suggested that a NFV arranger should be included in their platform with a high-density field layer, in future work. Therefore, it will be possible to test with the scaling-in/out capabilities of the system that may alter an additional sturdy operation in the industrial atmosphere.

A paper suggested by Ateya *et al.* (2017b) illustrated that there is a tendency to introduce a completely unique approach towards a cloud based structure, mostly on a cellular system, in which the little cell area units are connected with micro-cloud units with little capabilities to present the sting computing facilities. Additionally, the micro-cloud area units are connected to mini-cloud units that have higher capabilities. Also, the core network cloud connects the mini-clouds within the whole system. Introducing additional levels of cloud reduces the spherical trip latency and the network congestion. Moreover, the most vital challenge in realizing TI is the latency demand. Hence, so as to get a true time tactile communication and interaction, the end-to-end system latency should be the human latent period. TI needs an end-to-end latency of 1 ms; this includes all delays from the supply to the destination. Finally, the 1 ms latency also includes the process delay through the transmitter and receiver infrastructure hardware. In this paper, a structure class-conscious cloud-based system has been introduced for reducing latency and realizing the TI. Finally, there is a trade-off between using mini-clouds and their price. Thus, the optimization of the quantity of mini-clouds used in the network and therefore the range of micro-clouds connected to a mini-cloud is required.

Gupta *et al.* (2019) presented a paper that demonstrates the development of medical technology; the rising of 5G, TI, robot and computing technologies have enabled the knowledge domain innovations facilitating the occurrence of surgery technology. Moreover, within the medical field, the introduction of automation technology has contributed to telesurgery. Also, the telesurgery automation allotted with 5G TI infrastructure and artificial intelligence (AI) technology as the core fight will promote the audio, visual and tactile perceptions of a doctor throughout surgery and solve issues of resource scheduling. Therefore, this paper introduced a telesurgery automaton supporting the 5G tactile web and computing technology. Additionally, the design, composition, characteristics and advantages of telesurgery are explained from two aspects, the intelligent tactile feedback and human–machine interaction knowledge. On this basis, a human–machine interaction improvement theme throughout telesurgery is given from some aspects, i.e. edge-cloud integration, network slice and intelligent edge cloud. Finally, this paper has discussed the open problems with the given telesurgery system, concerning the ultra-high reliability, AI-enabled surgery automation, communication and security, to produce the reference for the promotion of the telesurgery automation performance. Finally, this paper has combined the 5G network, TI, AI and automation technology within the medical field, which introduces telesurgery automation to support a 5G tactile network.

Braun *et al.* (2017) suggested that future applications, such as driverless cars, industrial Internet and sensible grids can similarly demand high information measure, resilience, security and low latency communication. Those technical needs are met by the 5G communication system, which specializes in the longer-term air interface. Also, this may need minimum latency services that require advanced migration techniques, which are not simply fast. In this paper, the virtualization tools provide migration techniques, such as Docker and KVM that have been planned to examine and compare them. Additionally, an application-level migration protocol was proposed that excludes the drawbacks of the previous projects. Further, conjointly implementing the planned protocol in an exceedingly latency-sensitive gaming application was suggested. Moreover, considering that the server is transparently migrated from hosts users is evidence of conception, relinquishment is studied very well. This paper has analyzed the pros and cons of virtualization and containerization like KVM and Docker for server migration. Further, it shows that they are not the optimal selection for realizing mobile edge cloud. The breakdown of the migration time indicates that the state migration between the servers is within the order of tens of milliseconds once the servers are getting ready to contact one another, which will be the foremost common case within the MEC state of affairs. In conclusion, the paper has shown that MEC would be one of the most important components within future 5G networks to deliver low latencies, and demands the facilitation of sunshine weight.

9.2. 5G architecture design for Tactile Internet

In this section, we propose a design of an interoperable architecture for 5G and TI based on IEEE 1918.1, by mapping the functional modules of their architectures to each other. We will first start with the tactile edges and then the network domain.

9.2.1. *Tactile edge A*

The tactile edge produces haptic data, as well as other conventional Internet data. The production of application data depends on the use cases, which are identified by IEEE 1918.1 as follows (Aijaz *et al.* 2018):

1) "teleoperation;
2) automotive;
3) immersive virtual reality (IVR);
4) Internet of drones;
5) interpersonal communication;
6) live haptic-enabled broadcast;
7) cooperative automated driving".

As a part of the tactile edge, the GNC plays an important role in the interoperability of the tactile edge with any kind of network domain; all of the functionalities related to QoS are performed there. However, the design of the GNC was not described extensively in IEEE 1918.1, although all management and orchestration functionalities are implemented there. We therefore suggest a new design for the GNC that can be adapted to the different use cases. Indeed, the GNC can be considered as the interface between the master, slave and the network domain, as it is not only forwarding data to the network domain, but also applying some rules and mechanisms of QoS to guarantee the requirements of the different audio, video and haptic data generated in the human-to-human and/or human-to-machine communications. The challenge that should be addressed by the GNC, is how to satisfy the QoS constraints for each media type. Specifically, violating the haptic QoS constraints would destabilize the global control loop, leading to undesirable consequences on the ongoing applications (Al Ja'afreh *et al.* 2018). Indeed, the classical definition of QoS is associated with the ability of a network to provide the required services for the different types of traffic that are generated throughout the network. In the context of TI with a mixture of traffic (traditional and haptic data), which can be again classified as mission or non-mission critical, the primary goal is to provide priority with respect to their requirements of QoS. Hence, adding a method to distinguish among the application flows in the tactile edge will guarantee the required resources for the critical applications. Meanwhile, this differentiation requires constantly obtaining knowledge of the state of the network, so an appropriate decision regarding packet forwarding should be made (Luo *et al.* 2012; Zhou *et al.* 2016). Hence, we suggest the introduction of the following modules in the GNC, in order to intercept the flows and decide how to deal with them depending on their type. These modules are listed below with brief descriptions (see Figure 9.1):

– the **admission and congestion control** module is designed to decide whether or not to reject the requests from TDs, in case their QoS requirements are violated, and then send feedback status information to them;

– the **QoS mapping** module performs QoS mapping through a process of automatic translation between the representations of QoS flows and the QoS mechanisms for different technologies. This module is very important since the GNC has several interfaces; wired and wireless connecting the tactile edge to the network domain;

– the **resource monitoring** module makes use of the statistics of the current network states in terms of bandwidth, load, delays, etc. that are collected and maintained to make them available for the resource management module;

– the **resource management and optimization** module is responsible for managing resources efficiently, by providing E2E bandwidth guarantees, especially for those flows with higher priority. This is achieved by configuring the output queues of the GNC node, according to scheduling algorithms and policies that give the haptic data a higher priority compared to the other types of data.

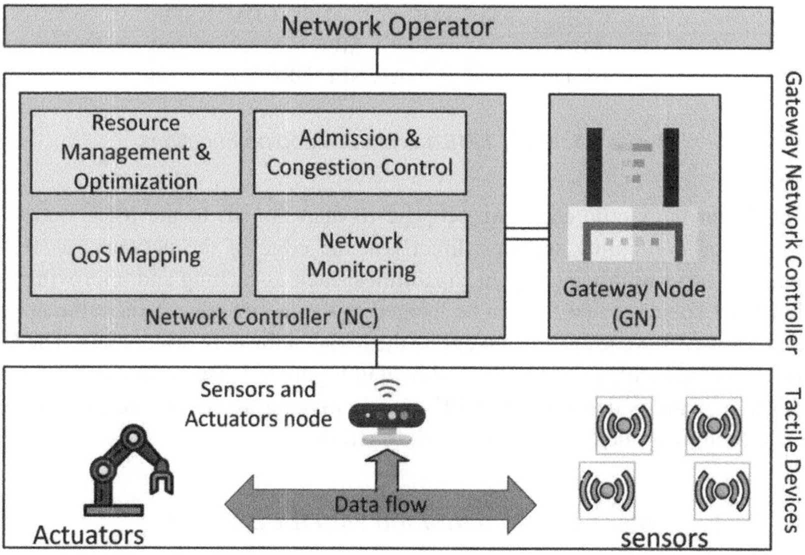

Figure 9.1. *GNC architecture. For a color version of this figure, see www.iste.co.uk/ali-yahiya/tactile.zip*

In our architecture, we propose a broad range of interfaces for the GNC (A, O and S interfaces) in order to connect different TDs to the network domain. Practically, it can have an interface to the network domain through the 5G NR network and another one through fiber optics. The latter is used for mission critical applications that require ultra-low latency, just like real-time teleoperations, while it can have WiFi or other interfaces connected with the TDs (T_b interface).

9.2.2. *Network domain*

In order to deliver TI services, we use 5G as the network domain starting from the RAN to the core network, as shown in Table 9.1. We map the functionalities of the UPE of the TI to UPF in 5G, as both are responsible for user plane functions, such as context activation, data forwarding to external networks and QoS support. The CPE is mapped to the corresponding modules in the 5G control plane, i.e. AMF, SMF, PCF, AUSF, UDM and NSSF, which are responsible for mobility management, policy and charging mechanism, and network slicing functionalities. The functionalities of the TSM are covered by the AF, NEF and NRF modules, since they are responsible for defining the characteristics of the services and exposing them to the tactile edges.

IEEE 1918.1	5G
UPE	UPF
CPE	AMF, SMF, PCF, AUSF, UDM, NSFF
SE	MEC
TSM	AF, NEF, NRF

Table 9.1. *IEEE 1918.1 and 5G Mapping Functions*

Finally, in the tactile edge, we propose to map the SE to the MEC entity that delivers cloud computing services directly at the edge of the network, as well as carrying out other highly intelligent capabilities, such as caching and improving QoE. IEEE 1918.1 specifies that SE can be located in the tactile edge A or in the network domain. However, we propose to integrate the MEC in the network domain. The MEC can be collocated with a local UPF, as shown in Figure 9.2. This is due to the fact that the 5G user plane is delegated to UPF, as it is playing a central role in routing the traffic to desired applications and network functions.

9.2.3. *Protocol stack of 5G integration with IEEE 1918.1*

The protocol stack that was adopted by IEEE 1918.1 is based on the TCP/IP model. Figure 9.3 shows both the user and control planes, as well as the connectivity between the layers starting from the TD to the core network of the 5G, which is our selected transport network for the haptic data. The user plane protocol stack starts from the TD that provides haptic, sensing and other types of data to the GNC. However, depending on the type of application generating the data, the application layer protocols will differ. For example, real-time protocol for interactive applications (RTP/I) can be adopted for classical Internet real-time applications, while haptic over IP (HoIP) can be used for haptic applications, where an adaptive sampling rate is needed. The GNC has a dual-protocol stack: one of them is based on the WiFi and/or Ethernet so that it can be connected to a network of TDs, and the other one is the 5G NR protocol stack that is connected to the gNB using NR via the T_b interface (3GPP 2018). Regarding the control plane, layers 1 and 2 of the protocol stack of the TD are the same as the ones used in the user plane; however, the third layer will be different since it is using the control plane protocol (CPP) to handle all of the functionalities related to the control plane, such as registration and authentication. Again, the key interoperability or connectivity between the tactile edge and the network domain is the GNC, which has two interfaces: one is used to be connected to the TD through the CPP, and the other one is represented by the 5G NR that connects the GNC to the gNB, as well as to the AMF, through the non-access-stratum (NAS) protocol, which provides the authentication, security, IP assignment, etc.

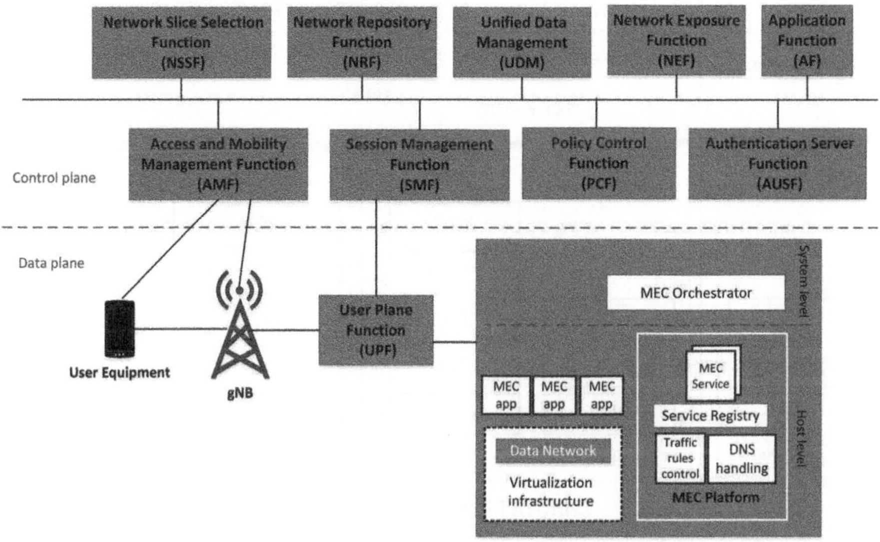

Figure 9.2. *MEC integrated with 5G. For a color version of this figure, see www.iste.co.uk/ali-yahiya/tactile.zip*

9.3. Haptics over IP

The HoIP can be described as a simple application-layer protocol, using UDP as the transport layer deployed in the network haptic devices. HoIP is significantly based on adaptive sampling and is designed to provide user flexibility in choosing different samples. HoIP software is developed with C++ and is designed to reduce the E2E delay between sender and receiver. Hence, round trip delay consists of haptic force capture and packetization delay at the transmitter, the receiver packet processing and rendering delay, and the network delay. Also, adaptive sampling is used with UDP depending on the perceptually essential features of the haptic signal, such as Weber's law and level crossings, and the reduced transmitted data is given to the administrator based on adaptive sampling at the haptic loop rate. Further, HoIP is originally designed to transmit haptic data over local area network (LAN). Figure 9.4 shows the TCP/IP stack and the position of HoIP.

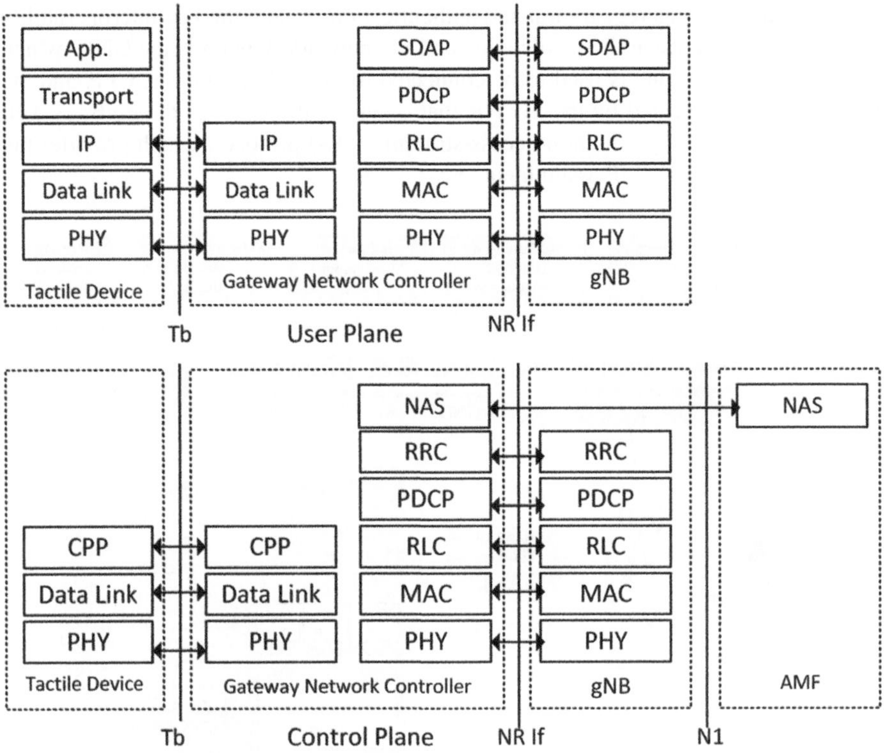

Figure 9.3. *User and control planes of TI and 5G integration*

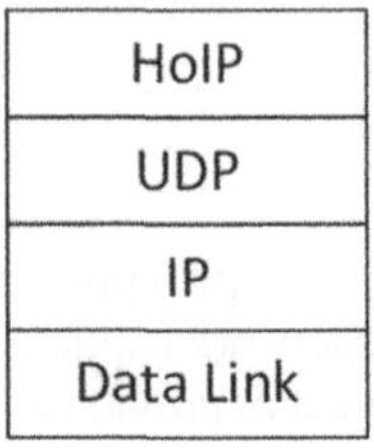

Figure 9.4. *HoIP in the protocol stack*

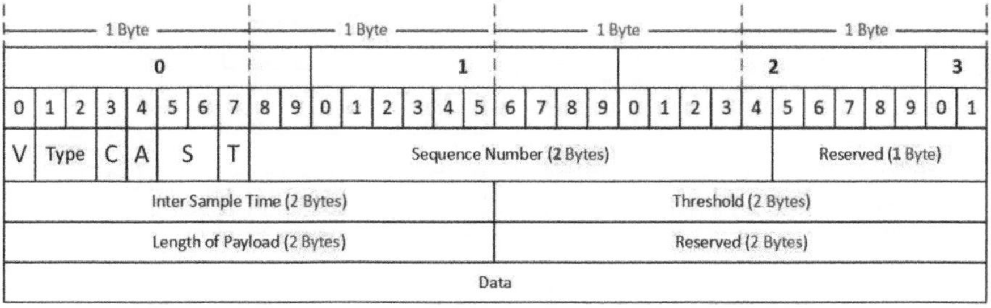

Figure 9.5. *HoIP in the protocol stack. For a color version of this figure, see www.iste.co.uk/ali-yahiya/tactile.zip*

9.4. Teleoperation case study

In this section, we focus on a teleoperation case study where a user can control a robotic hand remotely. More specifically, this is the case of an ungrounded haptic glove connected to a teleoperated robotic hand on a remote site. This can be used in applications where manual operation is needed in a hazardous remote environment, for example, as well as in the context of handling objects remotely (Kofman *et al.* 2005).

The three main parts of the TI architecture are involved in this application: the master and slave domains and the transport network. In the master and slave domains, both sensors and actuators are used to generate and transform data to realize a haptic control system. The types of sensors needed in the master domain are motion, position and force sensors, which can be built into a glove, as in Nishimura *et al.* (2014), Pacchierotti *et al.* (2017) and Perret and Vander Poorten (2018). Haptic actuators in the master domain give the kinesthetic force feedback from the remote robotic hand when handling objects. This is generated by using force sensors in the remote slave domain system. In addition to the sensors and actuators, a video signal is sent from the slave side for better dexterity when handling objects.

The E2E delay of haptic information, while being transferred through the network domain, is a critical operational parameter for a direct control closed-loop remote system, in contrast to a supervisory control teleoperator system, where a high E2E delay is tolerated for an autonomous controlled-remote system. Indeed, E2E delay of the order of 1–10 ms for a highly dynamic system is required to experience a realistic haptic effect, as well an average data rate of 1–4 k packets per second (kpps) (Aijaz *et al.* 2018). In other words, each packet should contain the sample information from each sensor and actuator using a sampling rate of 1–4 kHz.

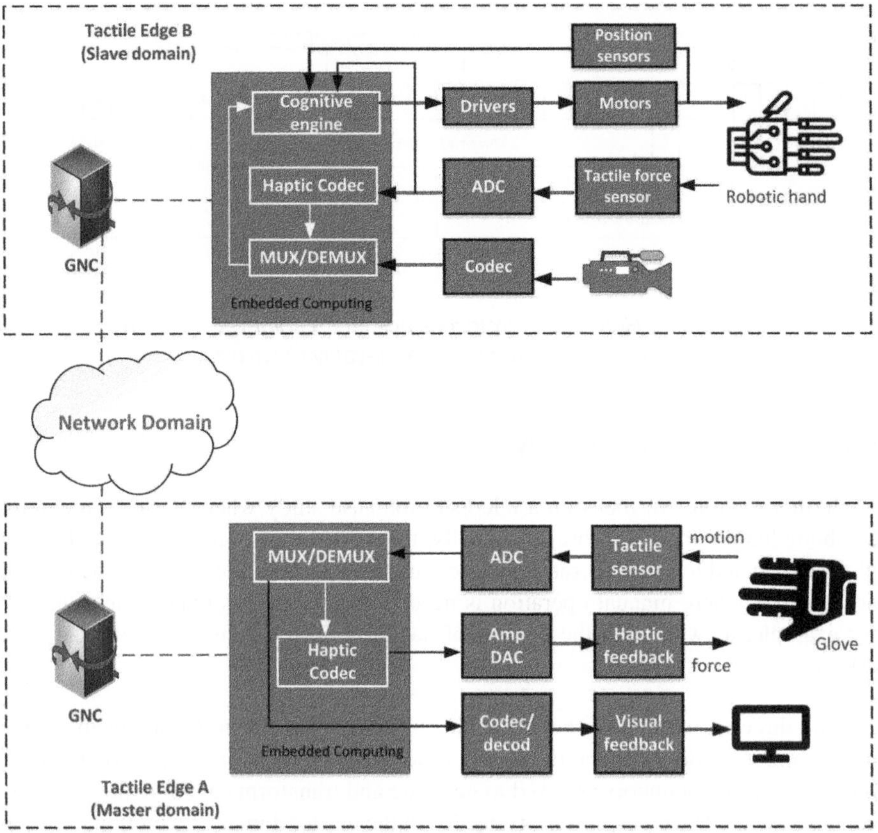

Figure 9.6. *Teleoperation system design. For a color version of this figure, see www.iste.co.uk/ali-yahiya/tactile.zip*

The architecture of the case study is shown in Figure 9.6, with a closed-loop control system on the slave side using the position and tactile force sensors. The haptic and visual feedback through the network are used to close the global control loop. For the case of a telehaptic application, closing the loop for a stable operation needs a 1 kHz haptic sampling rate. This means a packetization of 1,000 packet/s to be sent and received via the network.

To implement the suggested haptic remote control, the sensors and actuators are categorized as follows: five analogue bending sensors for the fingers, three gyroscopes to sense the orientation of the hand, six pressure sensors, one per fingertip and one for the palm of the hand. Concerning haptic feedback, we consider only using vibrators or piezoelectric devices on the tips and the palm of the hand to provide the sense of touching any object on the slave side (i.e. no force feedback is implemented since the glove has no actuators to do so). Instead, a video feedback signal augmented with haptic data coming from the slave TD edge is used to inform the operator to manipulate and grasp objects. The amplitude of the vibration can be used as a feedback indicator of the applied force by the robotic hand.

Indeed, the performance of our system is characterized by the transmission of all of the generated data with a minimum time delay. The 5G core network technology can guarantee short transmission delays, as explained in the suggested architecture. In the following sections, we will investigate the possible packetization of the data in the radio resource frame. We will only consider the uplink and downlink channels on the master side (i.e. the glove side in Figure 9.6), as the slave side will handle the same amount of data.

9.4.1. *Master to slave (uplink) data rate in edge A*

In this section, the amount of control and data bits from the master to the slave domain is found after digitizing the position and force data, in addition to a number of overhead bits. These bits are needed to differentiate the multiplexed data from the different sensors.

First, to find the amount of position data generated by the glove on the master side, we will fix the sampling of the sensors on the glove side to 1,000 Hz, which is faster than the speed of movement of a hand executing a thorough operation. Thus, we will have five analogue bending sensors on the glove, assuming eight bits ADC and three bits to distinguish between the five fingers' information; then, $1,000 \cdot 5 \cdot (8 + 3) = 55,000$ bits/sec are generated. Second, the hand orientation is sensed by the use of three gyroscopes, using 10 bit A/D converters and 1,000 Hz sampling rate, and two bits overhead to address the pitch, roll and yaw rotations; we then get: $1,000 \cdot 3 \cdot (10 + 2) = 36,000$ bits/sec. Finally, for the six force pressure sensors, there will be one per fingertip, plus one on the palm of the hand and three bits overhead to address the different sensors. A sampling frequency of 1,000 Hz will produce: $1,000 \cdot 6 \cdot (10 + 3) = 78,000$ bits/sec. According to the previous calculations, the total number of bits per second for the position orientation and force pressure data is then **169,000 bits/sec**. This gives a total of 169 bits/ms or 169 bits/packet for a sampling rate of 1 kHz.

9.4.2. *Slave to master (downlink) data rate in edge B*

On the slave TD edge side, we suggest video media augmented with force/pressure sensors to close the global control loop. For the six force/pressure sensors sampled at 1 kHz, 10 bits ADC and the three addressing bits, we have $1,000 \cdot 6 \cdot (10 + 3) = 78,000$ bits/sec. We use the the H.264 codec for video data compression. For an acceptable video quality, a data bit rate between 0.500 and

1.5 Mbps is possible (Tamhankar and Rao 2003). By considering the highest data bit rate of 1.5 Mbits/sec, a total of 1,500 kbps + 78 kbps = **1,578 kbits/sec** of raw data has to be sent from the slave side.

9.4.3. *Encapsulating the haptic data in HoIP*

Upon the multiplexing of the raw data, a haptic protocol should be used to insert some additional overheads, which is specific to haptic operations. To this end, we use the HoIP (Gokhale *et al.* 2013a, 2015). This protocol is more useful in the case of adaptive haptic sampling, where connection speed can be restricted.

Depending on the GNC type on the user plan (see Figure 9.8), the HoIP acts as the application layer, which in turn becomes the application layer of the 5G UE in our use case. Otherwise, if heterogeneous TD with different communication stacks are available, then the HoIP should be implemented on the top of the TD stack.

In order to find the number of overhead bits added to the previously calculated position and force/pressure data, we use the format suggested in Gokhale *et al.* (2013a). This encapsulates the data into segments and packets, and then consequently transport frames. The HoIP can be implemented in both the master and slave domains, and the following calculations show the amount of overhead added by the different layers, starting from the HoIP layer to the IP layer:

– Master to slave: starting from the application layer, the HoIP overhead per packet is found by adding the HoIP, UDP and IP. We can then obtain the number of overheads by adding 96 (HoIP) + 64 (UDP) + 192 (IP), giving 352 bits/packet. Summing both data and overhead, we obtain $169 + 352 = 521$ bits/packet (i.e. 521 kbps in the uplink direction).

– Slave to master: on this side, we have five force sensors for each of the robotic hand fingertips and one in the palm of the mechanical hand, in addition to the video signal. By using the same application layer protocol and a sampling rate of 1,000 samples/sec, we obtain $1,000.6.(10 + 3) = 78,000$ bps. A proportional–integral–derivative (PID) system can be used on the slave side to generate the haptic signal feedback to the operator side (master domain), before encapsulating it into the HoIP and then the UDP and IP layers. The total number of bits per packet (data+overhead) obtained from the slave side is: $78 + 1,500 + 352 = 1,930$ bits/ms or (1,930 bits/packet) on the uplink path of the slave side.

9.4.4. *5G network data and control handling*

In the previous sections, we found the data rates necessary for a correct functioning of the system. Here, we will check the possibility of transmitting these data rates on a 5G NR wireless network with an acceptable delay. This can be achieved by analyzing the frame structure of the UL and DL radio communications, since they have a significant impact on the QoS parameters, such as latency and throughput (Vihriala *et al.* 2016).

The NR interface has some new features like the massive multiple-input multiple-output (MIMO), small-size base stations, carrier aggregation (CA), beamforming and full-duplex. These evolutions are largely introduced in the physical layer (Zaidi *et al.* 2016). The NR is envisioned to operate from sub-1 GHz to 100 GHz to cover a variety of services and deployment options. It supports scalable OFDM numerology, with 2^n scaling of subcarrier spacing, which creates a family of OFDM waveforms with frame structures for the NR–air interface (Zaidi *et al.* 2016; Jeon 2018). This family is classified into FR1, which supports the subcarrier spacings of 15, 30 and 60 kHz, and FR2, which supports the subcarrier spacings of 60 and 120 kHz (Jeon 2018). The NR supports the QPSK, 16QAM, 64QAM and 256QAM modulation schemes for downlink and uplink transmissions (ETSI 2018d). Based on these physical layer parameters and the data rate estimated in the previous section, the feasibility of our use case can be determined. Assuming the configurations, as shown in Table 9.3, we can calculate the total data rate per subframe (where 1 subframe = 1 ms) that the radio access can handle as follows:

(Number of symbols per subslot · Number of subcarriers · Number of bits per OFDM symbol · Number of resource blocks (RB) · Number of slots per subframe)

which gives a total of:

$$14 \cdot 12 \cdot 4 \cdot 20 \cdot 2 \cdot 2 = 26,880 \text{ bits/subframe.}$$

Parameter	Value
Subcarrier separation	30 kHz
Carrier frequency/BW	3–6 GHz/10 MHz
SSB block length (L)	8
Number of resource blocks RB	20
Modulation scheme	16 QAM
MIMO	no MIMO
Number of slots per subframe	2

Table 9.2. *Table of 5G NR use case parameters*

The part of the synchronization signal block (SSB) used for the synchronization and cell search for the above case, can be found to be 6% of a subframe length (see Campos (2017)). Therefore, $0.94 * 26,880 = 25,536$ bits/subframe can be used for data and control for both UL and DL connections.

In fact, two types of information elements are exchanged through the radio links, namely cell search and synchronization control, and data and control. We are interested in finding out the overall generated bits. In the following, we will find out part of the radio link utilization:

– Synchronization and cell search: the NR supports a multi-beam operation using SSB, which is repeated L times to make an SS burst. The repetition of many bursts within 5 ms is called a burst set, which is used for a gNB beam-sweeping transmission. The periodicity of the burst set is of 20 ms (ETSI 2018e; Omri *et al.* 2019). The maximum number of L depends on the numerology used for the physical layer (ETSI 2018e). Considering the case with 15 kHz subcarrier spacing, $L = 4$ for carrier frequency $f < 3GHz$. This gives a total of 280 symbols for two frames, since the periodicity of a burst set is 20 ms. Knowing that the PSS, SSS and PBCH channels occupy four OFDM symbols, it can be found that only 90.0% of the radio resources remained for data and other control signals, while in the case of 30 kHz subcarrier spacing, with $L = 8$, the same radio resource availability for data and control is possible, but with a higher bit rate.

– Data and control: each IP packet found in section 9.10 (on the master and slave side) is considered as the service data unit (SDU) for the SDAP layer, which is then passed to the following sub-layers PDCP, RLC and MAC, as shown in Figure 9.3. Each layer will add a header (composed of some bytes) to the input SDU to obtain the protocol data units (PDU) on its output (Dahlman *et al.* 2018). Referring to the standardization documents ETSI (2018a), ETSI (2018b), ETSI (2018c) and ETSI (2018f), a total of approximately 60 bytes is counted as a total number of bytes to be added as headers for different control tasks. Hence, by adding the $60 \cdot 8 = 480$ bits/packet (control) to the 512 bits/packet and the 1,930 bits/packet for the uplink and downlink radio connection on the master side, respectively, 992 bits/packet on the UL and 2,410 bits/packet on the DL are expected.

Finally, considering the LDPC algorithm for the channel coding, with a code rate of $(n = 5, k = 1)$, this produces a raw bit rate of $5,107$ bits/subframe, to be compared with the sum of the uplink and downlink data and control bit rates (i.e. $2,410 + 992 = 3,502$ bit/subframe). Since the radio link bit rate (5,107 bits/subframe) is greater than the overall data, plus the control bit rate (3,502 bits/subframe), we can conclude that the 5G NR can easily handle the requirements of a haptic data communication, in terms of data rate. Since the overall data and control are contained in 1 ms duration (i.e. one subframe), this would guarantee a transmission delay within the required range for haptic perception.

9.4.5. *Case study operational states*

In this section, we introduce the operational states to show the steps from the establishment of a connection to the termination between the TDs. Based on the generic TI (device and architecture) operation states given in the IEEE 1918.1, we present a Moore finite state machine (FSM), along with a state mapping to the functional block in the proposed user and control plane architecture. The FSM illustrated in Figure 9.7 represents an E2E communication between two TD, through an edge–network–edge paradigm. This generally starts with a **registration** process, carried out by a TD within the tactile edge itself; however, this would be visible through the GNC, which should in turn, register to the services of the 5G through message exchange through the gNB with AMF, UPF and UDM, which are considered as the main modules of the core network that are involved in the process of registration. Therefore, the GNC can be considered as a special UE as it has the same capabilities of any UE, as well as the capability of relaying TDs to the network domain through its multiple interfaces. Then comes the **authentication** phase, where the AMF, AUSF and UDM functions will be involved. These are the modules that provide access, authentication and authorization, as well as the security credentials used during the session. Once registered and authenticated, the **control synchronization** would start between the GNC and the SMF for establishing an E2E connection with other TDs in other edges. In this phase, it should be decided which state the transition should go to, according to the type of data. If the data is haptic, then the next step will be **haptic synchronization**, where GNC and SMF are involved. These functional blocs are both responsible for setting up some parameters, such as codecs and session parameters, message formats, etc. In the case of the classical types of data, the **operation** state will be encompassed through GNC, MEC, UPF and DN. These functions are located in the data plane. During any data or control transmission, failure or errors may happen. Hence, we add two transitions to the **operation** state: one of them is an **operation recovery**, which indicates errors in the data and only involves GNC to manage or re-transmit the data. The second state is the **recovery**, which indicates a failure in the transmission, and uses GNC to fix it, and as a consequence, a transition will be made to the state of **control synchronization**, in order to enable the re-synchronization of the haptic data transmission. In both cases, once the recovery is performed, a transition from **operation** state to **termination** will be done to indicate a normal completion of the transmission.

9.4.6. *Case study protocol stack*

Figure 9.8 shows the protocol stack for our use case, with reference to Figure 9.3, which we have suggested for the integration of IEEE 1918.1 and 5G. Depending on the system design of the case study, in the user plane, there will be a direct communication between the application layer (the data multiplexing and packetization) of the TD and the GNC. The data multiplexer and demultiplexer is the part to be implemented on an embedded system. The HoIP layer can either be built in the GNC or in the TD, subject to the access method to the GNC. The UDP and IP layers will create the packets to be passed to the 5G stack layers in the GNC. After transmission of data and control through the radio, a similar stack on the gNB side will ensure the communication of data on the user plane 9.3. Concerning the control plane protocol stack, there will be no control messages generated by the tactile device; the GNC will take this in charge through the CPP module, which generates control messages regarding session registration, authentication, control synchronization, etc. following the transition states shown in Figure 9.7.

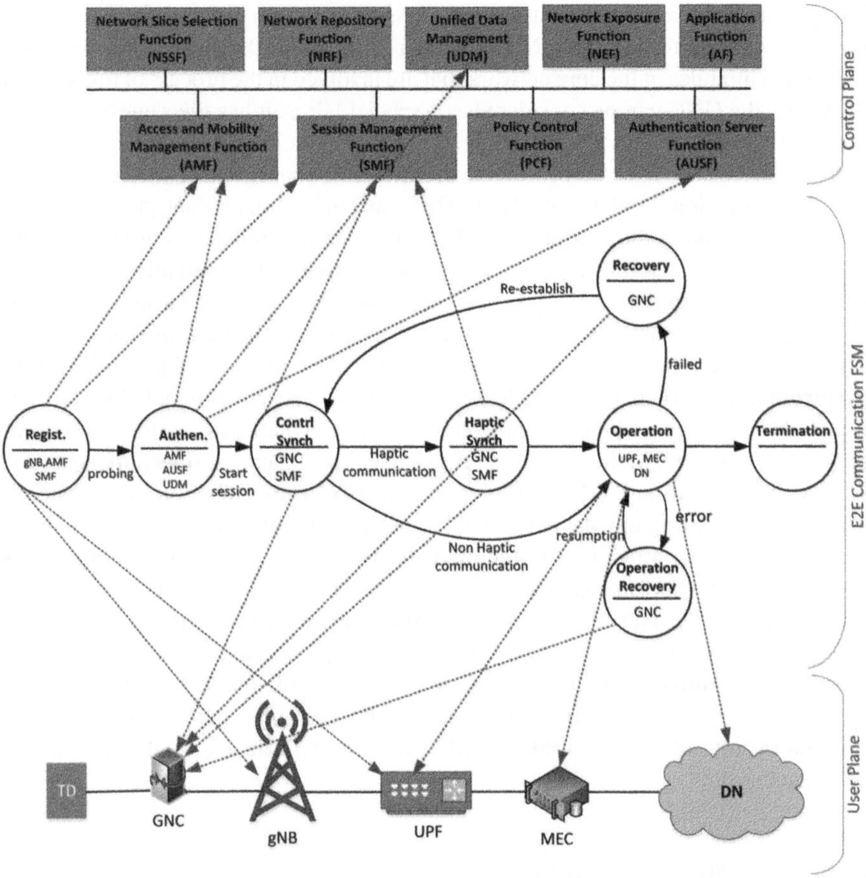

Figure 9.7. *Moore FSM for E2E communication in an integrated 5G and IEEE 1918.1 architecture. For a color version of this figure, see www.iste.co.uk/ali-yahiya/tactile.zip*

9.5. Simulation results

In this section, we use the network simulator-3 (NS3) to simulate our conceptual case study with a mmWave module implementing the 5G NR part (Mezzavilla *et al.*

2018). The parameters given in Table 9.3 are used for the physical layer. The default values of all of the other non-mentioned simulation parameters are taken for the simulations.

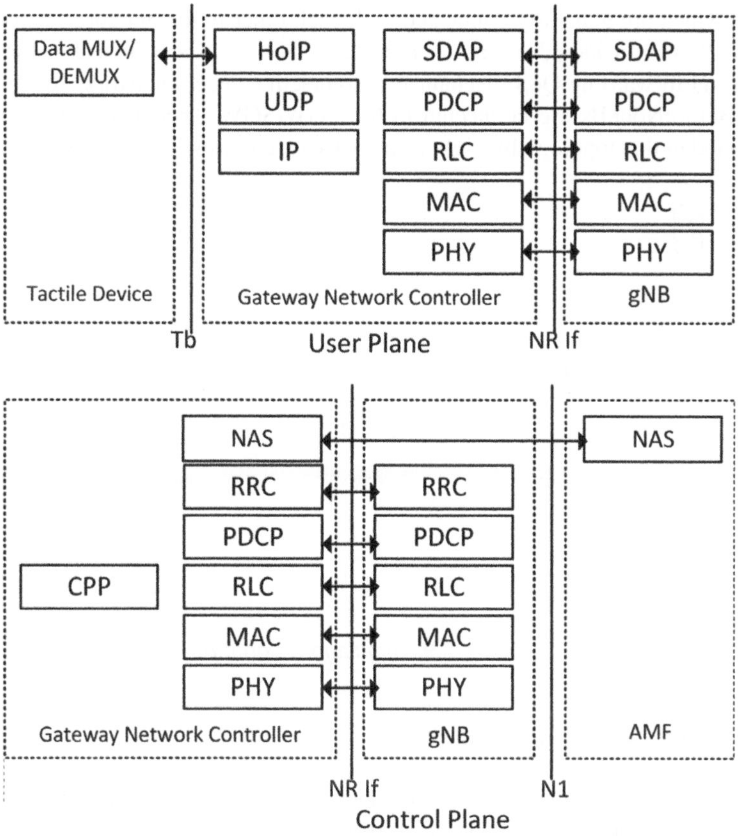

Figure 9.8. *User and control planes of the use case*

9.5.1. *Simulation topology*

A topology is defined for the considered 5G network architecture and is shown in Figure 4.2. The end-to-end network is represented by master and slave domains, connected together through the 5G core network as a network domain. The master and slave domains have the same capability and capacity in terms of network and node features. Each of them is composed of a 5G access network containing 25 stationary UEs, randomly distributed around the evolved node bases (eNodeBs) at both ends. The association with the eNodeB occurs according to the signal-to-interference-plus-noise ratio (SINR) values. Since the mmWave is used, its higher frequency does not permit the signal to travel long distances; hence, the distance set between UEs and eNodeB ranges between 30 and 150 meters. All UEs are generating haptic data and communicating with the eNodeB through the mmWave technology in 5G NR. Both access networks are connected via the packet gateway (PGW)/service gateway (SGW), which are generally collocated with each other in the 5G core network. The PGW/SGW is considered as the user plane in 5G, as it is responsible for packet routing and forwarding QoS mechanisms, connecting access networks together, and so on. Both eNodeBs are connected to the PGW/SGW through wired technology, represented in this topology by high-speed network cables with 100 Gbps data rate.

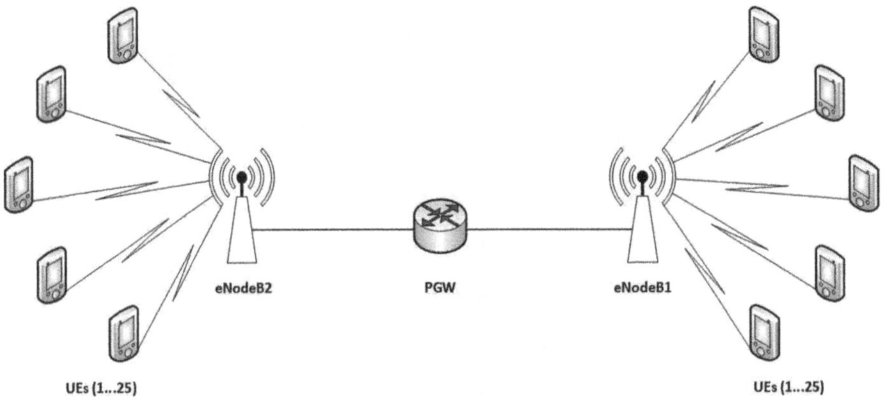

Figure 9.9. *Simulation scenario*

9.5.2. *NS3 network architecture*

The NS3 is considered as a full-stack simulator, as it models all of the network-layer protocol stacks, starting from the physical layer to the application layer. For each layer, different models are used, depending on the functionalities enabled in that layer. However, in order to technically implement these models through the C++ programming language, a modular technique for simulating the network models is used. In this thesis, the mmWave module is used to simulate the end-to-end communication system, where the main difference with LTE is in the physical layer. Figure 9.12 is proposed for the end-to-end network architecture, including all of the modules used in the different network layers.

We start with the UE and its protocol stack model. In the higher layer represented by the application layer, the HoIP is used to generate haptic data. In order to model HoIP, the packet sink application module is used to generate packets, respecting the same format of HoIP as in Gokhale *et al.* (2013a). Then, the generated data will be encapsulated in UDP by creating a client application that sends packets adding 64 bits as overhead. Afterwards, the packets will be encapsulated in the IP layer by adding its header information with a 192 bit overhead. This would give a total header for the packet, including header of HoIP + header of UDP + header of IP. The rationale behind these values can be based on the original values of the header of the IP (Osanaiye and Dlodlo 2015) and UDP (Long and Zhenkai 2010), shown in Figures 9.10 and 9.11, respectively.

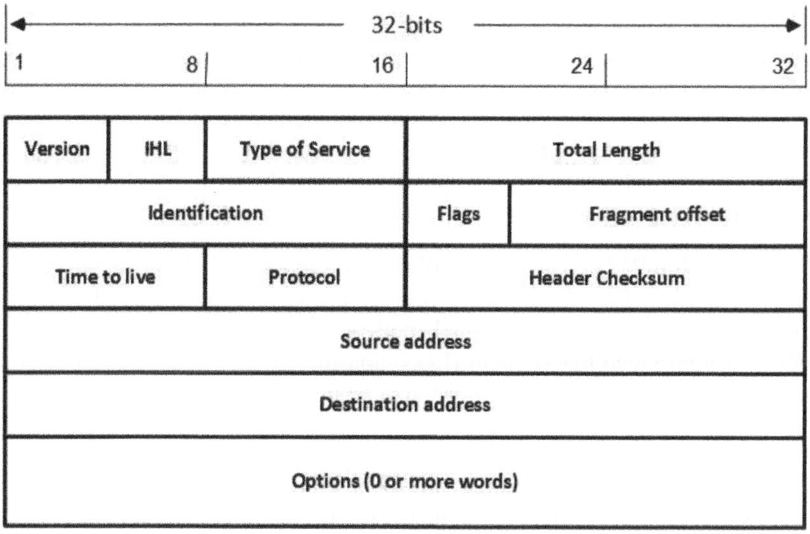

Figure 9.10. *IP header*

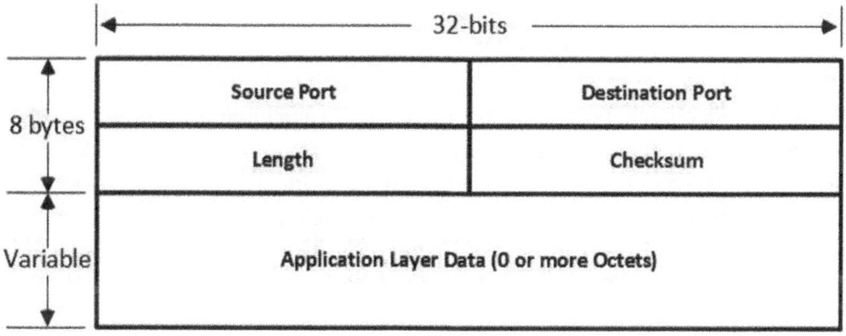

Figure 9.11. *UDP header*

Since the UE is connected to the eNodeB through the 5G NR interface, they will have the same radio stack, where mmWave module is installed. The NetDevice classes supporting the 5G NR interface used in both eNodeB and UE, are

MmWaveEnbNetDevice and MmWaveUeNetDevice, respectively. Both NetDevices will include PHY and medium access control (MAC) layer functionalities as follows:

– MmWaveEnbMac and MmWaveUeMac: These classes are used to provide the MAC mechanisms used for accessing the channel, as well as deploying the method of re-transmissions. The adaptive modulation and coding (AMC) scheme is used here, with interactions with the physical layer, is based on information provided by the MiErrorModel, especially the SINR, while HARQ is used to enable the re-transmission of the erroneous packets, using the soft combining method. This would allow the error correction procedure when errors occur during transmission. The class scheduler exists only in the eNodeB implementation of mmWave, since it is the centralized entity that controls the communication among UEs. Four schedulers were implemented in the eNodeB just like the Round Robin, Proportional Fair, Earliest Deadline First and Maximum Rate schedulers. In this thesis, Round Robin is used to allocate the resources in terms of slots to the UEs.

– MmWaveEnbPYH: The 5G mmWave is supporting TDD; therefore, TDD frames are used in the simulation. Since the physical layer is directly connected to the channel model, several channel models were used, including the path loss model, which provided line of sight (LOS)/non-line of sight (NLOS) characteristics and small-scale fading, while beamforming and multiple antennas were used to gain capacity in the network.

In the proposed architecture (as shown in Figure 9.12), the eNodeB has two protocol stacks as it has two network interfaces. The first one uses 5G NR to connect to the UEs, and the other is connected to the 5G core network through PGW/SGW through a network cable. This is the reason behind seeing two protocol stacks in the eNodeB node architecture. On the top of the radio stack is the EpcEnbApplication, which is responsible for bridging the data generated in the data plane from the UE and data plane part from PGW/PGW. Note that the GTP or GPRS tunneling protocol (GTP) is used to decapsulate packets coming from the eNodeB and tunnel them to the PGW/SGW (since packets do not have the same format) that it is connected to. Then, the PGW/SGW will encapsulate the payload to be transmitted to the eNodeB2 through the EpcSgwPgwApp class; thus, the eNodeB2 will receive the packet and send it to the UE destination. Note that the path crossed by packets in the UL direction is similar to the aforementioned description, but in the opposite direction.

9.5.3. *Simulation scenario*

Two different scenarios are studied in this thesis. As shown in Figure 4.2, the main difference is the load generated in each one of them. In the first scenario, the HoIP is used in the application layer and encapsulated in the UDP datagram. Then, an IP packet composed of 1,400 will be transmitted over 5G NR in each frame (i.e. 1 ms); this scenario can be named as a low-load scenario. In the second scenario, 4,200 bytes are transmitted, which is triple the size of the packet used in the same scenario. The aim is to study the performance of the HoIP with two different loads in two different channel models for path loss, namely line of sight (LoS) and non-line of sight (NLOS).

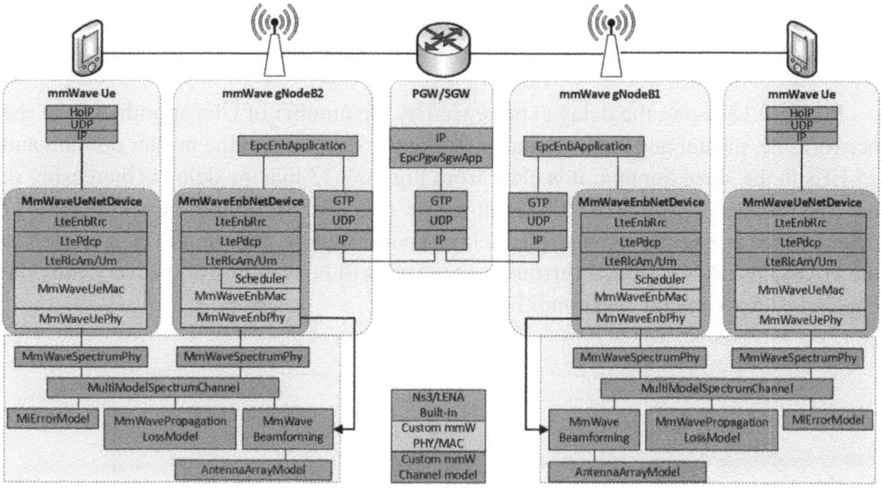

Figure 9.12. *End-to-end network architecture. For a color version of this figure, see www.iste.co.uk/ali-yahiya/tactile.zip*

Parameter	Value
Number of UE	50
Number of eNB	2
Subframes per frame	10
Subframe length	100
Symbols per subframe	24
Symbol length	4.16
Number of subbands	72
Subband width	13.89
Subcarriers per subband	48
Center freq	28 GHz
NumDlCtrlSymbols	1
NumUlCtrlSymbols	1
NumHarqProcesses	10
MaxPacketSize	1,400, 4,200
maxPacketCount	50
interPacketInterval	1,000μs
NumHarqProcess	20

Table 9.3. *Simulation parameters*

9.5.4. *Simulation results*

9.5.4.1. *Low-load scenario*

The most important performance parameters that were studied in this chapter are delay and throughput. Both parameters were studied in the case of LoS and NLoS models of the channel.

Figure 9.13 shows the delay experienced by the number of UEs at both ends of the network, i.e. master and slave domains, as there are 25 UEs in the master domain and 25 UEs in the slave domain. It is clear from Figure 9.13 that the delay is increasing in the case of NLoS compared to LoS; however, it is not violating the QoS in terms of delay, which is fixed to 10 ms. The delay experienced by UEs in NLoS is related to the HARQ retransmission, as erroneous packets will be re-transmitted every time the channel is experiencing bad conditions.

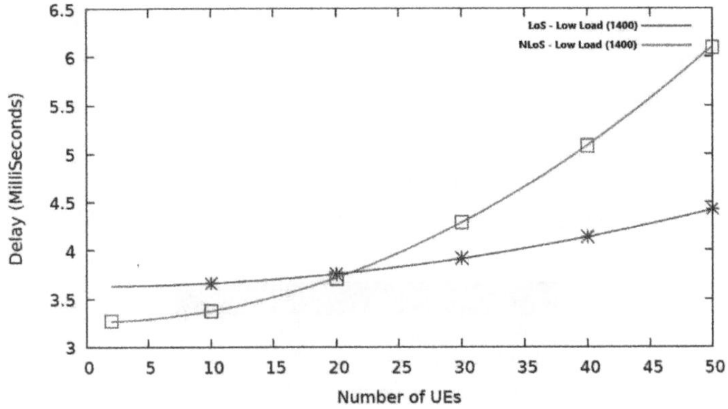

Figure 9.13. *Delay versus number of UEs for low load (LoS and NLoS). For a color version of this figure, see www.iste.co.uk/ali-yahiya/tactile.zip*

Figure 9.14 shows the throughput in the case of low load, while LoS and NLoS are considered. The throughput in the case of LoS is higher than in the case of NLoS. This is a normal behavior since the channel is not experiencing any bad conditions. However, for the same amount of data to be transmitted, there will be errors and some packets that will not be fully transmitted because of channel conditions.

9.5.4.2. *High-load scenario*

Again, Figure 9.15 illustrates the delay in the high-load scenario, where the packet size is 4,200 bytes for both channel models LoS and NLoS. The delay in the case of NLoS is higher than in the case of LoS. The delay is increased in the case of NLoS, since the channel varies over time and its quality changes accordingly. In the case of an erroneous channel, the HARQ mechanism will be used to correct the errors; however, it demands some redundancy in the retransmission, according to the implemented mechanism of retransmission. This would increase the E2E delay, as shown in Figure 9.15. In the simulation, the number of HARQ retransmissions is fixed to 20, which would increase the delay, as the delay includes not only the transmission of the correct data, but also data with errors through the HARQ mechanism. It is noted that when the number of users reaches 40, there will be so many errors in the packets that no packets will transmitted anymore. Hence, the curve will stop at 40 UEs.

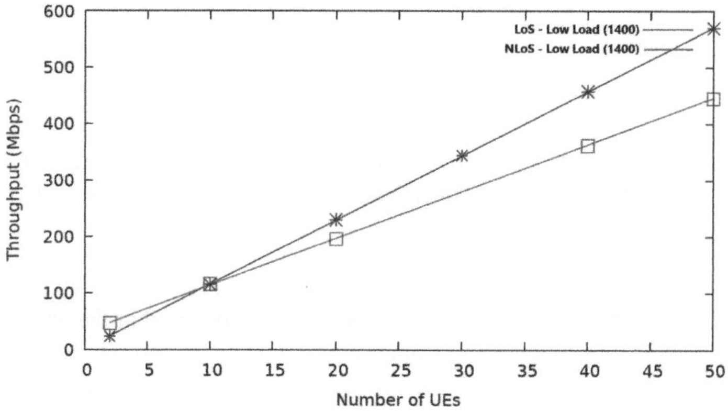

Figure 9.14. *End-to-end network architecture. For a color version of this figure, see www.iste.co.uk/ali-yahiya/tactile.zip*

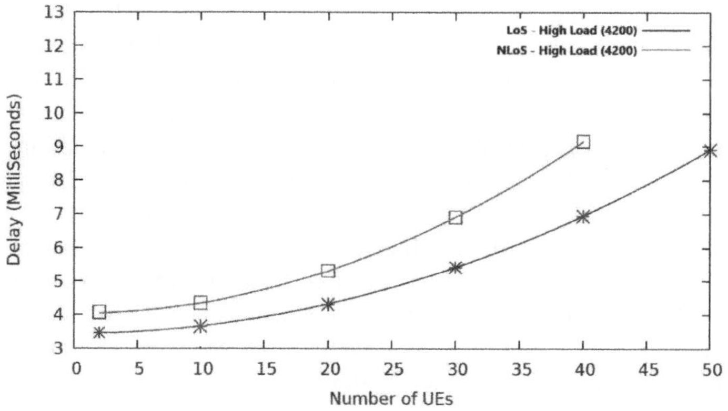

Figure 9.15. *Delay versus number of UEs for high load (LoS and NLoS). For a color version of this figure, see www.iste.co.uk/ali-yahiya/tactile.zip*

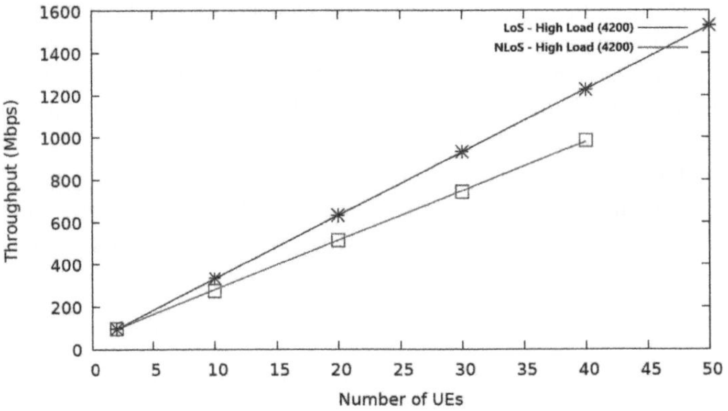

Figure 9.16. *Throughput versus number of UEs for high load (LoS and NLoS). For a color version of this figure, see www.iste.co.uk/ali-yahiya/tactile.zip*

As shown in Figure 9.16, the throughput in the case of LoS for high load is higher than in the case of NLoS. Again, the same issue which occured in the previous figure, occurs here. This means that the channel condition is bad and when the number of users is increases, the number of big-size packets will increase. This means that there are more errors and simulation will stop at this stage, due to the huge number of HARQ retransmission. It cannot go beyond 20 re-transmissions.

9.5.4.3. *Comparison between low-load and high-load scenarios*

Figure 9.17 shows the delay in both cases when high and low loads are considered under the LoS channel condition. The delay of LoS includes the delay of the transmission, plus the delay that the packets are experiencing in the PGW/SGW. This is due to the technology difference of communication, as when frames are transmitted between the gNodeB1 and PGW/SGW, the MTU of both technologies are different. The frames transmitted from the UE and gNodeB1 are transmitted respecting the mmWave format of frame; then, when it is transmitted to the PGW/PGW through the high-speed Ethernet, the MTU will change. In this case, fragmentation will occur in PGW/SGW as the LTE frames are bigger than the MTU of Ethernet, which is 1,500 bytes. Thus, the delay of fragmentation will be added to the delay of transmission. The same process will occur, in inverse, between the PGW/SGW, as the frames should be defragmented to be transmitted to the gNodeB2.

The delay comparison of low load and high load under the NLoS condition is considered in Figure 9.18. The same concern encountered in the previous figure, regarding the delay, is true; however, the delay of transmission of HARQ will be added, so here, the delay especially consists of the high load to propagation delay + transmission delay + HARQ delay + fragmentation/defragmentation delay.

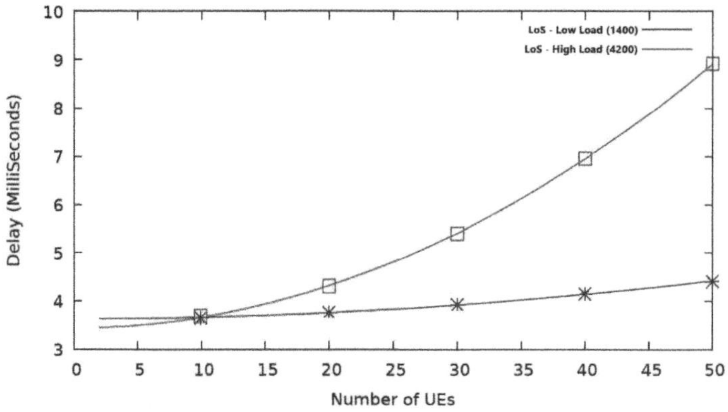

Figure 9.17. *Delay versus number of UEs for LoS (low load and high load). For a color version of this figure, see www.iste.co.uk/ali-yahiya/tactile.zip*

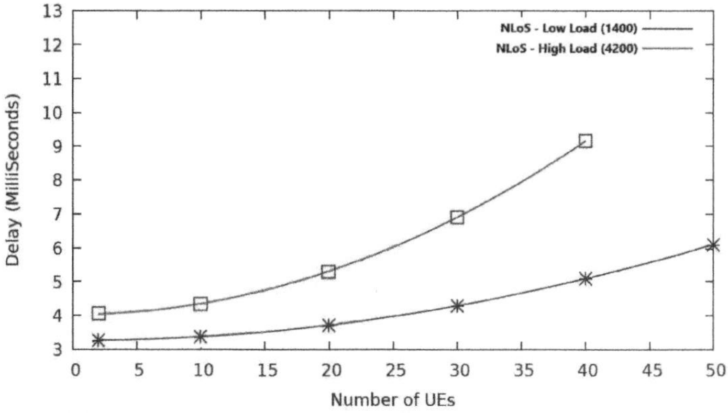

Figure 9.18. *Delay versus number of UEs for NLoS (low load and high load). For a color version of this figure, see www.iste.co.uk/ali-yahiya/tactile.zip*

Throughput is compared between high load and low load under LoS and NLoS scenarios. It is obvious that throughput in the case of high load under LoS is higher than in the case of low load, as shown in Figure 9.19. This is because the impact of the size of the packet is so large.

In Figure 9.20, the throughput in the case of LoS is higher for high load compared to low load, until the number of UEs reaches 40. When the number of UEs are greater than 40, errors are generated in the channel, such that the channel is no longer able to transfer data.

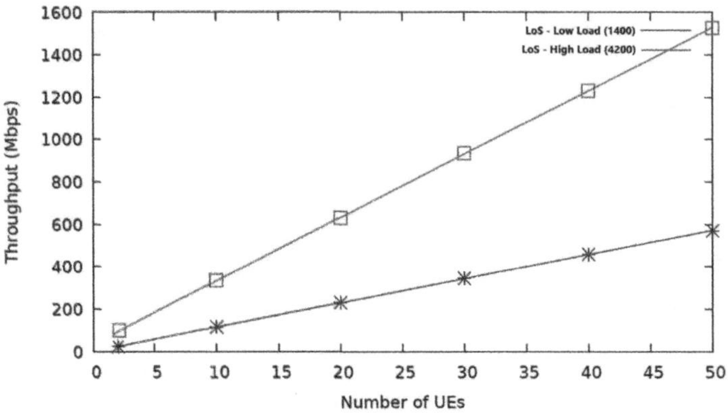

Figure 9.19. *Throughput versus number of UEs for LoS (low load and high load). For a color version of this figure, see www.iste.co.uk/ali-yahiya/tactile.zip*

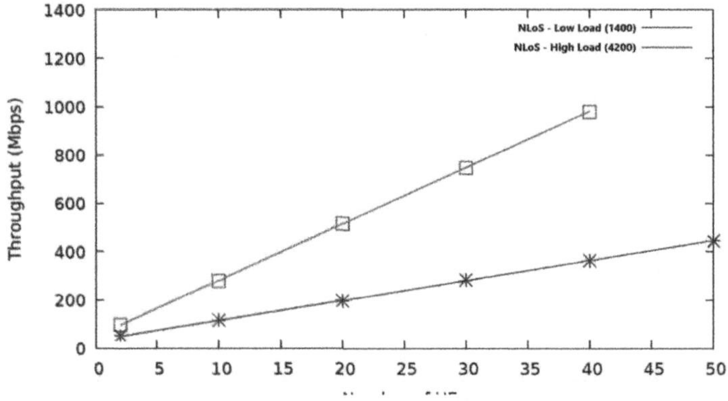

Figure 9.20. *Throughput versus number of UEs for NLoS (low load and high load). For a color version of this figure, see www.iste.co.uk/ali-yahiya/tactile.zip*

9.5.4.4. *Channel quality versus delay and throughput*

The channel quality represented by the path loss model, with the option of NLOS, is used to investigate the QoS performance in low and high loads in terms of delay and throughput. Tables 9.4 and 9.5 show the throughput and the delay versus the average SINR for a transmission over a period of 12 seconds, to average the fading effects of the channel. The path loss was increased while keeping the UE position fixed. As the SINR decreases, the MAC will adapt its modulation scheme's MCS level to encode the data. As shown, the throughput in the high load is higher than the low load, since the data packet size is higher. While the delay in both cases is not affected by the channel quality, the difference in delay is in the order of fractions of milliseconds, which can be considered as jittering.

Loss	SINR (dB)	MSC	Delay (ms)	Throughput (Mbps)
80	134.06672606	25	4.0917941	229,337
100	25.77653649	25	3.951734	229,342
120	13.18842548	18	3.4127152	229,361

Table 9.4. *SINR versus delay and throughput for NLoS (low load – 1,400)*

Loss	SINR (dB)	MSC	Delay (ms)	Throughput (Mbps)
80	34.38363847	25	4.4823337	666,763
100	25.67782306	25	5.0857908	666,632
120	12.9915719	18	4.9344621	637,401

Table 9.5. *SINR versus delay and throughput for NLoS (high load – 4,200)*

9.6. Conclusion

It is expected that 5G, with its new radio interface, will bring changes in the landscape of communications. This is due to its support to the vertical applications that require very low latency with ultra-reliability. In parallel, the IEEE 1918.1 standard for TI is in progress to deliver different use cases and applications. In this chapter, we proposed a conceptual telehaptic application, with numerical analysis of the data rate to be transmitted over the 5G NR, taking the case of mmWave-spectrum band. Using the HoIP with its related additional overhead of the multimodal data, we found that the haptic data and the 5G control headers can be easily embedded in a subframe carrier of the NR in 1 ms, and the 5G network architecture will then ensure the 1 ms transmission delay required for haptic signals. We showed that for a single user, despite a degraded channel quality, the data rate and delay requirements are still ensured.

9.7. References

3GPP (2018). NR and NG-RAN overall description; Stage 2. Technical Specification (TS), 3rd Generation Partnership Project.

Aijaz, A., Dohler, M., Aghvami, A.H., Friderikos, V., Frodigh, M. (2017). Realizing the tactile internet: Haptic communications over next generation 5G cellular networks. *IEEE Wireless Communications*, 24(2), 82–89.

Aijaz, A., Dawy, Z., Pappas, N., Simsek, M., Oteafy, S., Holland, O. (2018). Toward a tactile internet reference architecture: Vision and progress of the IEEE P1918.1 standard. *CoRR* [Online]. Available at: http://arxiv.org/abs/1807.11915.

Al Ja'afreh, M., Adharni, H., El Saddik, A. (2018). Experimental QoS optimization for haptic communication over tactile internet. *IEEE International Symposium on Haptic, Audio and Visual Environments and Games (HAVE)*, 1–6.

Antonakoglou, K., Xu, X., Steinbach, E., Mahmoodi, T., Dohler, M. (2018). Toward haptic communications over the 5G tactile internet. *IEEE Communications Surveys Tutorials*, 20(4), 3034–3059.

Ateya, A.A., Muthanna, A., Gudkova, I., Vybornova, A., Koucheryavy, A. (2017a). Intelligent core network for tactile internet system. *Proceedings of the International Conference on Future Networks and Distributed Systems*, ACM Press [Online]. Available at: https://doi.org/10.1145/3231053.3231125.

Ateya, A.A., Vybornova, A., Kirichek, R., Koucheryavy, A. (2017b). Multilevel cloud based tactile internet system. *19th International Conference on Advanced Communication Technology (ICACT)*, 105–110.

Ateya, A.A., Muthanna, A., Gudkova, I., Abuarqoub, A., Vybornova, A., Koucheryavy, A. (2018). Development of intelligent core network for tactile internet and future smart systems. *Journal of Sensor and Actuator Networks*, 7(1), 1.

Braun, P.J., Pandi, S., Schmoll, R., Fitzek, F.H.P. (2017). On the study and deployment of mobile edge cloud for tactile internet using a 5G gaming application. *14th IEEE Annual Consumer Communications Networking Conference (CCNC)*, 154–159.

Campos, J. (2017). Understanding the 5G NR physical layer. *KEYSIGHT Technologies*. Santa Rosa, CA.

Cen, Z., Mutka, M.W., Zhu, D., Xi, N. (2005). Supermedia transport for teleoperations over overlay networks. *International Conference on Research in Networking*. Springer, Berlin, 1409–1412.

Dahlman, E., Parkvall, S., Skold, J. (2018). *5G NR: The Next Generation Wireless Access Technology*. Elsevier Science, London.

Dodeller, S. (2004). Transport layer protocols for haptic virtual environments. PhD Thesis, University of Ottawa, Canada.

Dohler, M., Mahmoodi, T., Lema, M.A., Condoluci, M., Sardis, F., Antonakoglou, K., Aghvami, H. (2017). Internet of skills, where robotics meets AI, 5G and the tactile internet. *European Conference on Networks and Communications (EuCNC)*, 1–5.

Eid, M., Cha, J., El Saddik, A. (2011). Admux: An adapative multiplexer for haptic audio visual data communication. *IEEE Transactions on Instrumentation and Measurement*, 60(1), 21–31.

ETSI (2018a). 5G; NR; radio link control (RLC) protocol specification (3GPP TS 38.322 version 15.3.0 release 15), Technical Specification (TS), ETSI.

ETSI (2018b). 5G; NR; medium access control (MAC) protocol specification (3GPP TS 38.321 version 15.3.0 release 15), Technical Specification (TS), ETSI.

ETSI (2018c). 5G; NR; packet data convergence protocol (PDCP) specification (3GPP TS 38.323 version 15.2.0 release 15), Technical Specification (TS), ETSI.

ETSI (2018d). 5G; NR; physical layer; general description (3GPP TS 38.201 version 15.0.0 release 15), Technical Specification (TS), ETSI.

ETSI (2018e). 5G; NR; physical layer; general description (3GPP TS 38.201 version 15.0.0 release 15), Technical Specification (TS), ETSI.

ETSI (2018f). LTE; 5G; evolved universal terrestrial radio access (E-UTRA) and NR; service data adaptation protocol (SDAP) specification (3GPP TS 37.324 version 15.1.0 release 15), Technical Specification (TS), ETSI.

Gokhale, V., Dabeer, O., Chaudhuri, S. (2013). Hoip: Haptics over internet protocol. *IEEE International Symposium on Haptic Audio Visual Environments and Games (HAVE)*, 45–50.

Gokhale, V., Chaudhuri, S., Dabeer, O. (2015). Hoip: A point-to-point haptic data communication protocol and its evaluation. *Twenty First National Conference on Communications (NCC)*, 1–6.

Gupta, R., Tanwar, S., Tyagi, S., Kumar, N. (2019). Tactile-internet-based telesurgery system for healthcare 4.0: An architecture, research challenges, and future directions. *IEEE Network*, 33(6), 22–29.

Holland, O., Steinbach, E., Prasad, R.V., Liu, Q., Dawy, Z., Aijaz, A., Pappas, N., Chandra, K., Rao, V.S., Oteafy, S., Eid, M., Luden, M., Bhardwaj, A., Liu, X., Sachs, J., Araújo, J. (2019). The IEEE 1918.1 "tactile internet" standards working group and its standards. *Proceedings of the IEEE*, 107(2), 256–279.

Jeon, J. (2018). NR wide bandwidth operations. *IEEE Communications Magazine*, 56(3), 42–46.

King, H.H., Tadano, K., Donlin, R., Friedman, D., Lum, M.J.H., Asch, V., Wang, C., Kawashima, K., Hannaford, B. (2009). Preliminary protocol for interoperable telesurgery. *International Conference on Advanced Robotics*, 1–6.

King, H.H., Hannaford, B., Kammerly, J., Steinbachy, E. (2010). Establishing multimodal telepresence sessions using the session initiation protocol (SIP) and advanced haptic codecs. *IEEE Haptics Symposium*, 321–325.

Kofman, J., Xianghai Wu, Luu, T.J., Verma, S. (2005). Teleoperation of a robot manipulator using a vision-based human-robot interface. *IEEE Transactions on Industrial Electronics*, 52(5), 1206–1219.

Li, C., Li, C., Hosseini, K., Lee, S.B., Jiang, J., Chen, W., Horn, G., Ji, T., Smee, J.E., Li, J. (2019). 5G-based systems design for tactile internet. *Proceedings of the IEEE*, 107(2), 307–324.

Long, W. and Zhenkai, W. (2010). Performance analysis of reliable dynamic buffer UDP over wireless networks. *Second International Conference on Computer Modeling and Simulation*, 1, 114–117.

Luo, T., Tan, H., Quek, T.Q.S. (2012). Sensor openflow: Enabling software-defined wireless sensor networks. *IEEE Communications Letters*, 16(11), 1896–1899.

Maier, M., Ebrahimzadeh, A., Chowdhury, M. (2018). The tactile internet: Automation or augmentation of the human? *IEEE Access*, 6, 41607–41618.

Mekikis, P., Ramantas, K., Antonopoulos, A., Kartsakli, E., Sanabria-Russo, L., Serra, J., Pubill, D., Verikoukis, C. (2020). NFV-enabled experimental platform for 5G tactile internet support in industrial environments. *IEEE Transactions on Industrial Informatics*, 16(3), 1895–1903.

Mezzavilla, M., Zhang, M., Polese, M., Ford, R., Dutta, S., Rangan, S., Zorzi, M. (2018). End-to-end simulation of 5G mmWave networks. *IEEE Communications Surveys Tutorials*, 20(3), 2237–2263.

Miao, Y., Jiang, Y., Peng, L., Hossain, M.S., Muhammad, G. (2018). Telesurgery robot based on 5G tactile internet. *Mobile Networks and Applications*, 23(6), 1645–1654.

Nasir, Q. and Khalil, E. (2012). Perception based adaptive haptic communication protocol (PAHCP). *International Conference on Computer Systems and Industrial Informatics*, 1–6.

Nishimura, N., Leonardis, D., Solazzi, M., Frisoli, A., Kajimoto, H. (2014). Wearable encounter-type haptic device with 2-DoF motion and vibration for presentation of friction. *IEEE Haptics Symposium (HAPTICS)*, 303–306.

Omri, A., Shaqfeh, M., Ali, A., Alnuweiri, H. (2019). Synchronization procedure in 5G NR systems. *IEEE Access*, 7, 41286–41295.

Osanaiye, O.A. and Dlodlo, M. (2015). TCP/IP header classification for detecting spoofed DDoS attack in cloud environment. *IEEE EUROCON 2015 – International Conference on Computer as a Tool (EUROCON)*, 1–6.

Osman, H.A., Eid, M., Iglesias, R., Saddik, A.E. (2007). Alphan: Application layer protocol for haptic networking. *IEEE International Workshop on Haptic, Audio and Visual Environments and Games*, 96–101.

Oteafy, S.M.A. and Hassanein, H.S. (2019). Leveraging tactile internet cognizance and operation via iot and edge technologies. *Proceedings of the IEEE*, 107(2), 364–375.

Pacchierotti, C., Sinclair, S., Solazzi, M., Frisoli, A., Hayward, V., Prattichizzo, D. (2017). Wearable haptic systems for the fingertip and the hand: Taxonomy, review, and perspectives. *IEEE Transactions on Haptics*, 10(4), 580–600.

Perret, J. and Vander Poorten, E. (2018). Touching virtual reality: A review of haptic gloves. *ACTUATOR 2018; 16th International Conference on New Actuators*, 1–5.

Rosenberg, J., Schulzrinne, H., Camarillo, G., Johnston, A., Peterson, J., Sparks, R., Handley, M., Schooler, E. (2002). SIP: Session initiation protocol. Report, RFC Editor, RFC 3261.

Schulzrinne, H. and Casner, S. (2003). RTP profile for audio and video conferences with minimal control. Report, RFC Editor, RFC 3551, STD 65.

Shirmohammadi, S. and Georganas, N.D. (2001). An end-to-end communication architecture for collaborative virtual environments. *Computer Networks*, 35(2–3), 351–367.

Szabo, D., Gulyas, A., Fitzek, F.H.P., Lucani, D.E. (2015). Towards the tactile internet: Decreasing communication latency with network coding and software defined networking. *Proceedings of European Wireless; 21th European Wireless Conference*, 1–6.

Tamhankar, A. and Rao, K.R. (2003). An overview of h.264/mpeg-4 part 10. *Proceedings EC-VIP-MC 2003; 4th EURASIP Conference Focused on Video/Image Processing and Multimedia Communications (IEEE Cat. No.03EX667)*, 1, 1–51.

Van Den Berg, D., Glans, R., De Koning, D., Kuipers, F.A., Lugtenburg, J., Polachan, K., Venkata, P.T., Singh, C., Turkovic, B., Van Wijk, B. (2017). Challenges in haptic communications over the tactile internet. *IEEE Access*, 5, 23502–23518.

Vihriala, J., Cassiau, N., Luo, J., Li, Y., Qi, Y., Svensson, T., Zaidi, A., Pajukoski, K., Miao, H. (2016). Frame structure design for future millimetre wave mobile radio access. *IEEE Globecom Workshops (GC Wkshps)*, 1–6.

Wongwirat, O., Ohara, S., Chotikakamthorn, N. (2005). An adaptive buffer control using moving average smoothing technique for haptic media synchronization. *IEEE International Conference on Systems, Man and Cybernetics*, 2, 1334–1340.

Zaidi, A.A., Baldemair, R., Tullberg, H., Bjorkegren, H., Sundstrom, L., Medbo, J., Kilinc, C., Da Silva, I. (2016). Waveform and numerology to support 5G services and requirements. *IEEE Communications Magazine*, 54(11), 90–98.

Zhou, J., Jiang, H., Wu, J., Wu, L., Zhu, C., Li, W. (2016). SDN-based application framework for wireless sensor and actor networks. *IEEE Access*, 4, 1583–1594.

10

Issues and Challenges Facing Low Latency in the Tactile Internet

Tara ALI-YAHIYA

Department of Computer Science, University of Paris-Saclay, France

10.1. Introduction

The TI has given rise to a wide range of use cases with diverse type of applications. However, they are not equal in terms of their needs for network resources, and there is consequently a diversity in the requirement of end-to-end QoS assurance needs. The new applications, especially the haptic applications, may need 1 to 10 ms as an end-to-end latency especially for teleoperation case studies. The network domain should support new infrastructure to allow ultra-low latency and high reliability. Let us consider the classical view of the actual mobile network and the traditional communication protocols. They are not suitable for the applications of TI as delay is introduced in every layer of the protocol stack in the end devices, the master domain, the transport domain for the end-to-end communication. Hence, 5G is considered as the perfect candidate for transporting haptic data due to the flexible design of its access and core networks.

Involving a user in real haptic communication would rather involve the other types of their interactive senses, just like involving the voice and the video as a supportive application. However, the decision to have a standalone codec for each type of data

The Tactile Internet,
coordinated by Tara ALI-YAHIYA and Wrya MONNET.

or a hybrid codec is under study as, when dealing with QoS, each type of traffic will require a specified QoS to be guaranteed through the network. Thus, a multiplexer interacts actively with each type of traffic depending how urgent the need for that type of traffic to be prioritized in terms of resource allocation (Aijaz *et al.* 2018; Sachs *et al.* 2019).

From a technical standpoint, the E2E latency may be introduced when the packet is traveling form one edge to another. For example, the multiplexing technique should take into consideration the high update and packet rate of haptic data compared to audio and video traffic. As it should also change its behavior not only when multiplexing and sending the data through the network, but also for congestion control when it happens, the multiplexer should decide which packet to drop and which has a higher priority to be passed through the channel (Eid *et al.* 2011; Gokhale *et al.* 2017a; Cabrera *et al.* 2019). The methodology used in the multiplexing algorithm would affect the latency to a great extent, and designing a good multiplexer is a challenging issue due to the variety of the type of Internet traffic including the haptic traffic.

The other hurdle in TI is the size of packets that influence the delay of transmission, as considering the current size would be a big obstacle towards providing the 1 ms end-to-end delay. This delay includes packet processing in the receiver and transmitter including encoding and decoding. The orthogonal frequency division multiplexing (OFDM) used in the current technology where the symbol duration is long may not be a good option for modulation.

As for accessing cloud computing capabilities, the delay will increase hugely if the cloud computing is accessed through the traditional Internet, this will bring a heavy load on the backhaul and core network, and will also increase the latency, thus violating all the threshold values for latency. Therefore, moving the capabilities of the cloud close to the edge would be a good solution, which means that processing and storage should be achieved in RAN and not done in the core network, i.e. the Internet, consequently reducing the delay to some extent.

Since the traffic model in TI does not follow a regular basis, i.e. the arrival of packets to the schedulers can be sporadic and/or bursty in the medium access control (MAC) layer in the entities involved in the access network, the delay experienced by a packet thus includes the transmission delay and queuing delay in the transmitter side, the decoding and delay in the receiver side. For the applications which have a hard latency requirement, the design of the queue delay model should be considered in the whole system. In view of the aforementioned factors impacting the delay of haptic applications, in this chapter, we describe the latest research works making use of different technologies to find a solution for guaranteeing the E2E delay.

10.1.1. *Technical requirements for the TI*

The technical requirements for the TI include the following:

1) Ultra-responsive connectivity: most TI applications need the end-to-end latency/round trip delay to be in the order of about 1 ms. The end-to-end latency refers to the summation of the transmission times needed while sending the information from a sensor/device or human for the case of haptic communication through the communication infrastructure to a control server, the information processing time at the server, the processing at variable communication hops (i.e. routers and switches) and the retransmission times through the communication infrastructure back to the end device or human.

2) Ultra-reliable connectivity: another vital need for the TI is ultra-reliable network connectivity, in which reliability refers to the probability of ensuring the necessary performance under presented system limits and conditions over a certain time interval. For example, the factory automation scenario in a smart factory requires a reliability of about 99.999% for about 1 ms latency. One of the solutions to improve the reliability for TI applications is to employ concurrent connections with many links, and to use many paths for graph connectivity to be aware of a single point of failure. Nevertheless, this approach depends on the channel dynamics and the availability of channel state information (CSI) knowledge. Having higher signal-to-noise ratio (SNR) margins in the link budget and using stronger channel coding are significant solutions towards improving the reliability of a communication link. Enhancing the reliability will assist in decreasing the latency because of the lower number of resulting retransmissions.

3) Distributed edge intelligence: proper artificial intelligence (AI)/ML techniques need to be examined to be used at the edge-side of the wireless TI networks in order to facilitate the interpolation/extrapolation of human activities and predictive caching for decreasing the end-to-end latency. Moreover, AI/ML-based predictive actuation methods need to be examined in order to augment the coverage of tactile services/applications.

4) Transmission and processing of tactile data: to simplify the transmission of tactile information over the packet-switched networks, tactile encoding mechanisms need to be improved. To handle the highly multidimensional nature of human tactile perception, an efficient audio/visual sensory feedback mechanism needs to be examined.

5) Security and privacy: other key requirements of the TI are security and privacy under latency restricts. To meet these requirements, physical layer security techniques with low computational overhead, secure coding techniques, and reliable and low-latency methods to identify the legitimate receivers should be analyzed (Sharma *et al.* 2020).

10.2. Low latency in the Tactile Internet

The TI is still in its infancy and new methods should be developed to guarantee the low-latency criteria that are one of the essentials among the others characterized by ultra-reliability, security and high QoS and QoE guarantee. However, when referring to the solutions that try to guarantee the minimum latency characterized by 1 ms for the critical haptic applications to multiple of 100 ms for other types of data combined with haptic data, the adaptive solution can be the best choice. Such kinds of solutions should adapt themselves intelligently to the context. To date, there have been many research works trying to deal with the low latency by keeping it as minimal as possible. The method used for achieving this objective can range from framework, resource allocation, technological methods like MEC, network coding and communication protocols. In the following, we detail some solutions used to reduce the latency, explaining the methodology used to achieve this.

10.2.1. *Resource allocation*

One of the most significant techniques used to reduce the latency is resource allocation. The main layers involved in resource allocation, and which would impact on the E2E latency in the protocol stack, are the physical (PHY) and MAC layers. A cross-layer design involving both layers can be considered as an efficient solution towards obtaining a reduced delay. One of the characteristics of the TI application is the bursty nature of the generated traffic. According to the packet arrival process, a burst of packet transmission involves a large amount of data sent over a short time. To exploit the burstiness, Hou *et al.* (2018) classified the packet arrival process into two states – high and low – and they designed different transmission techniques for both states taking into consideration the awareness of the base station (BS) with regard to the quantity of traffic sent by the users. Accordingly, the BS would classify both states based on the Neyman–Pearson method. The behavior of the BS will change according to the amount of traffic; it will reserve dedicated bandwidth for users with a high amount of traffic giving them higher priority, while for those in a low traffic state, a resource pool is shared.

In order to ensure the low E2E delay communication, She *et al.* (2016a) considered a short frame structure for transmission and took into consideration different parameters affecting packet loss during the transmission. Latency is bounded to delay of transmission and queue delay, while reliability is bounded to packet error probability, queueing delay violation probability and packet dropping probability. The authors worked on a cross-layer design of the MAC and PHY layers in order optimize these probabilities in relation to the power allocation in the BS. A proactive packet dropping mechanism is proposed to satisfy the QoS requirement with the limited transmit power. She and Yang (2016) use the basis of the work of She *et al.* (2016a) in order to use the solution in the context of vehicle collision avoidance; however, they optimized the bandwidth allocation for users based on the queue delay and the power allocation in order to ensure the reliability.

Again, Gholipoor *et al.* (2018) proposed a cross-layer framework that combines traditional and TI data so it can be more realistic. However, instead of using OFDM, they used sparse code multiple access (SCMA) that 5G proposed for a transmission paradigm. The aim of this chapter is to increase the sum rate of the traditional data while guaranteeing the delay of the TI. They proposed a queue in the transmitter to differentiate the TI from traditional networks, as well as in the receiver for which the BS in their cases proposed different codebooks and power allocations for both types of traffic.

She *et al.* (2016b) investigate the impact of spatial diversity and frequency diversity in ensuring the transmission reliability, and the total bandwidth required for a wireless system to support the QoS requirements of massive machine type devices. They employed a two-state transmission model to characterize the transmission reliability constraint based on the achievable rate with finite blocklength channel codes. They assigned multiple subchannels to each active device, from which the device simply selects one subchannel with the channel power exceeding a threshold for transmission after channel probing. They optimized the number of subchannels, the bandwidth of each subchannel and the threshold for each device to minimize the total bandwidth required by the system to ensure the reliability.

In Aijaz (2016), the authors used LTE-A networks to allocate resources in terms of resource blocks (RB) to the haptic devices involved in the communication. Joint uplink (UL) and downlink (DL) scheduling necessitates an information exchange mechanism between the UL and DL schedulers. They investigated the problem of radio resource allocation for haptic communications over 5G LTE-A networks. The radio resource allocation requirements of haptic communications, together with the constraints of UL and DL multiple access schemes, have been translated into a power and RB allocation problem. To reduce the complexity, the problem is first decomposed using an optimal power control policy and then transformed into a binary integer programming (BIP) problem for RB allocation. The authors proposed a low-complexity heuristic algorithm for joint UL and DL scheduling that not only fulfils the utility requirements of haptic devices, but also outperforms the classical algorithms.

10.2.2. *Mobile edge computing*

The MEC plays a very important role in the reduction of low latency and it is designed for this purpose. For example, in Ateya *et al.* (2017), the authors proposed a multilevel hierarchy of cloud units in order to reduce the round-trip delay in the mobile network, especially the LTE network, by introducing a new level of cloud units with higher capabilities in the path between the core network and eNBs so that the cloud unit (microcloud) reduces the communication latency. The authors used the concept of MEC in order to bring the processing function in the vicinity of users by introducing small micro and mini clouds in each level of the network in the path to the core network which in its turn can include the functionality of the cloud.

Rimal *et al.* (2017) proposed an MEC design on an integrated fiber-wireless (FiWi) access network. They used a hybrid architecture that integrates MEC and the conventional cloud for different types of applications depending on the nature of the application, whether it is latency sensitive or not. The author claimed that using MEC over FiWi improves the network performance in terms of QoS for different types of applications. In Saddik *et al.* (2011), the authors proposed to replay haptic content using a caching technique, which is useful for applications relying on haptic feedback. The caching mechanism is very useful in the case of haptic applications, as it is movement/behavior repetition, which means that there is no need to reproduce the same movement/behavior all the time, since they will be cached and reused whenever there is a need to produce them.

10.2.3. *Network coding*

Network coding is a method that is used to reduce the latency in the network and increase the communication efficiency, and especially the error probability, when sending data on an unreliable channel. This is done through network coding based on algebraic algorithms on each node that receive the data and then recode it and send it to the destination to be decoded. Hence, both nodes should be synchronized and the behavior of the network will depend on one hop communication rather than the end-to-end communication used in the traditional networks in which each node will store and forward, but not process, the data (Swamy *et al.* 2016).

With the introduction of SDN in 5G, the network coding can reduce the latency more through multihop networks by including its functionality in the controller. Hence, Szabo *et al.* (2015) used network coding, especially the random linear network coding integrated with SDN, in order to reduce latency in a multi-hop environment. They introduced the functionality of compute-and-forward to the coding via the use of SDN instead of routers using only the functionality of store-and-forward. This would improve latency and reduce packet re-transmission with respect to other traditional approaches.

10.2.4. *Haptic communication protocols*

In order to support haptic communication, either new protocols should be designed for the variable nature of the haptic traffic as it is a hybrid scheme, or the traditional network protocol should be modified to be adapted to it. Talking about the higher layers of the protocol stack, just like the transport layer, there are two classical options, namely the user datagram protocol (UDP) and the transmission control protocol (TCP). However, both protocols have their advantages and disadvantages when supporting haptic data transport. For example, TCP can be considered a heavy protocol and needs the connection to be established between the peers in order for a transmission to occur. With regard to reliability, TCP can be considered as a reliable protocol for haptic communication, but in terms of providing low latency, it would introduce extra latency which cannot be suitable in a tactile environment where low latency is required. Although UDP is a light protocol and most suitable for haptic communication, it does not meet the reliability requirement.

A protocol called Supermedia TRansport for teleoperations over Overlay Networks (STRON) (Saddik *et al.* 2011) was created to operate over overlay networks transmitting data using different network paths. STRON was compared against TCP and Stream Control Transmission Protocol (SCTP), showing that it performs significantly better in the case of a network that includes paths with heavy packet loss. Timely execution of the protocol handler tasks with real-time interrupts allows for more immediate transmission of haptic data packets. Furthermore, the efficient transport protocol (ETP) (Wirz *et al.* 2008, pp. 3–12) aims to reduce round-trip delay time which is related to the inter-packet gap (IPG). By monitoring the transfer rate, it is possible to optimize IPG by setting it to a minimum value in order to maintain stability and maximum performance of the haptic application (Cen *et al.* 2005, pp. 1409–1412; Wirz *et al.* 2008, pp. 3–12).

As for the application layer, temporal management is important since data can be haptic mixed with audio and video. The challenge is to aggregate all these types of traffic to be transported in one single data stream. Hence, synchronization between both ends is important and should be guaranteed. Almost all solutions proposed to multiplex these types of traffic using different algorithms in the literature would count on the UDP protocol as a light protocol (Gokhale *et al.* 2015, 2013).

Generally, the application layer protocols supporting audio, video and haptic modalities and according to Gokhale *et al.* (2017b) can be classified into:

(i) constant bit rate-based telehaptic protocols in which CBR data streams were injected into the network at a steady rate. Some examples of this protocol that adopted this method, albeit in a modified way, could include the application layer protocol for haptic networking (ALPHAN) (Osman *et al.* 2007), adaptive multiplexer (ADMUX) (Eid *et al.* 2011), Haptics over Internet Protocol (HoIP) (Nasir and Khalil 2012);

(ii) adaptive sampling-based telehaptic protocols in which haptic signal samples are injected into the network only taking part of whole samples into consideration due to the high bandwidth that they require, which is why in order to avoid this problem, only some samples will be identified and be sent (Clarke *et al.* 2006; Sakr *et al.* 2011; Bhardwaj *et al.* 2013).

10.3. Intelligence and the Tactile Internet

The work on progress in the developments of 6G, artificial intelligence (AI) and TI use cases will be leading towards what we call intelligent connectivity from an end-to-end point of view. This would include intelligence in how the operation will be performed in the master domain, network domain and slave domain. In fact, the network should adapt itself to the type of application without the intervention of a human through automatic or autonomic functionalities provided in each protocol stack of the network. All the use cases of TI will be easy to deploy thanks to the sophisticated mechanism introduced by artificial intelligence (Khemiri 2015; Cheng *et al.* 2018; Dab *et al.* 2019; Holland *et al.* 2019; Pérez *et al.* 2020; Yu *et al.* 2020; Zhao *et al.* 2020).

In order to achieve this intelligent connectivity, contextual data from the environment needs to be analyzed and processed to make the right decision at the right moment. This would reduce give alerts in real-time leading to fewer human errors. This can also be extended to the prediction mechanisms offered by AI through its different techniques just like machine learning techniques.

In this section, we will show how AI can intervene in all the domains of TI and how the TI can benefit from AI techniques in order to enhance the performance of the system especially for mission-critical applications that require very fast feedback from the slave domain. Indeed, IEEE 1918.1 introduced the reference architecture of TI by adding elements that would be good contenders for implementing AI techniques, taking the GNC and the support engine, for example. These elements could play a great role towards the zero-delay perception in the network depending on the type of closed or open-loop systems.

10.4. Edge intelligent

The support engine can play the role of edge computing in the TI, and it can reside either in the master domain or in the network domain depending on how the information can be uploaded or downloaded from it. The implementation of the role of processing and computation in the support engine can be supported by the AI mechanism in order to provide the so-called intelligent edge.

In Ahmad *et al.* (2021), the authors proposed to minimize the path length or to stretch the path between the user and the cached content when the content has been identified on which router it is placed. They introduced a reinforcement learning-based approach to reduce the stretch between the user and the content router. Therefore, their contributions were a step ahead by implementing AI in the information-centric networking to learn based on exploration and exploitation to give better performance. The idea is to decrease the delay from an end-to-end perspective while decreasing the number of paths in the network through the Q-learning-based approach that is the technique of reinforcement learning to reduce the stretch between the user and content router.

In Grasso *et al.* (2021), the authors implemented the AI technique in the tactile support engine as its aim is to predict the haptic/tactile experience, for example acceleration of movement on one end and force feedback on the other end, in order to send forecasted values to the controlled domain when needed, allowing the spatial decoupling of the active and reactive end(s) of the TI. The authors tried again to reduce the delay in a mission critical application while using on-line gaming.

In Monnet and Yahiya (2020), the authors introduced a teleoperation mechanism overlaid on 5G in order to guarantee the QoS of the end-to-end network. The AI technique was used to exchange information among MEC servers through the gossip protocol to have a global view of the whole network instead of having a local view. The work can be summarized by implementing the gossip protocol in the MEC server in order to exchange information with its peers as shown in Figures 10.1 and 10.2.

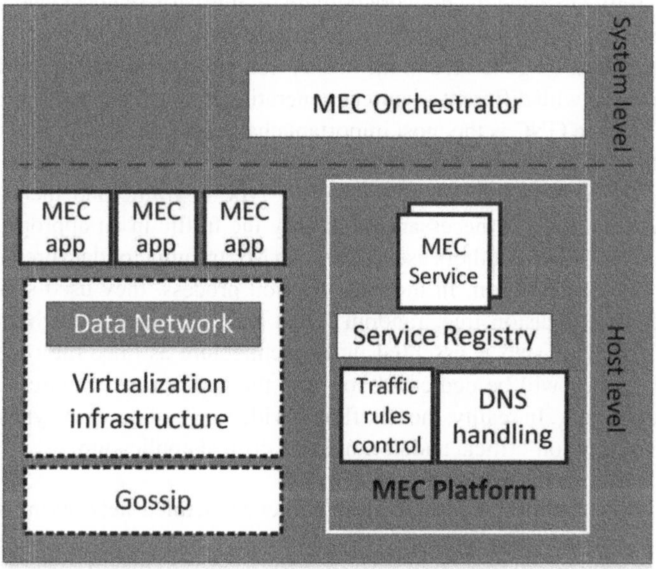

Figure 10.1. *MEC node-based gossip protocol. For a color version of this figure, see www.iste.co.uk/ali-yahiya/tactile.zip*

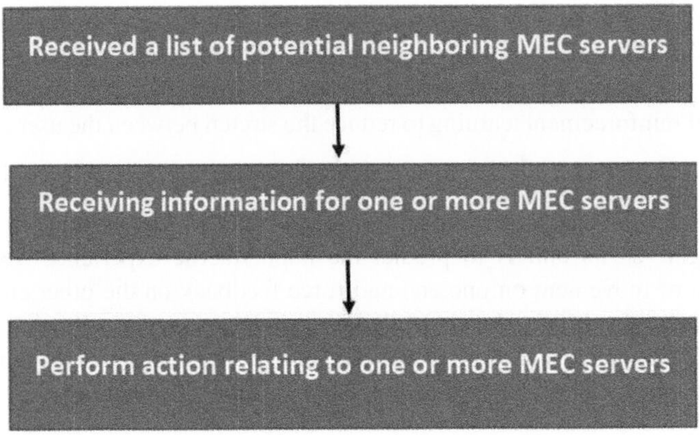

Figure 10.2. *Gossip protocol. For a color version of this figure, see www.iste.co.uk/ali-yahiya/tactile.zip*

However, the AI is not used only to guarantee the QoS over the network, it has other purposes just like traffic classification. In fact, AI techniques can be deployed to classify the traffic through the network domain. In this case, the GNC which plays the role of interfacing between master domain and network domain, can act as the anchor of traffic classification generated in the master domain. This is due to the diverse type of traffic in the TI with different resources generating traffic dependent certainly on the use case. Thus, the GNC is the most important challenge to deal with especially when dealing with applications requiring a stringent quality of service guarantee, especially in terms of latency. In order to recognize the type of traffic and then allocate the appropriate resource, it is necessary to classify the traffic in an appropriate way. In Amaral *et al.* (2016), the authors used SDN with ML training for data preparation, data clustering and classification. In the classification process, they used support vector machine (SVM), decision tree, random forest and Kth-nearest neighbor as shown in Figure 10.3. This step is essential in TI architecture as once the types of traffic are classified, they will be dedicated not only the appropriate resource but also the appropriate network. In reality, the TI offers a wide range of network types depending on the type of mission-critical and non-mission critical application.

Sheng (2018) proposed a scalable intelligence-enabled networking platform to remove the traffic redundancy in 5G audio–visual TI scenarios. The proposed platform incorporates a control plane, a user plane, an intelligent management plane and an intelligence-enabled plane. Out of these planes, the intelligence-enabled plane comprises a novel learning system that has decision-making capability for generalization and personalization in the presence of conflicting, imbalanced and partial data. Furthermore, Ruan and Wong (2018) studied the application of ML intelligence in taking effective decisions for dynamically allocating the frequency resources in a heterogeneous fiber-wireless network. More specifically, they investigated the utilization of an artificial neural network for the following purposes: (i) in learning network uplink latency performance by using diverse network features; and (ii) in taking flexible bandwidth allocation decisions towards reducing the uplink latency.

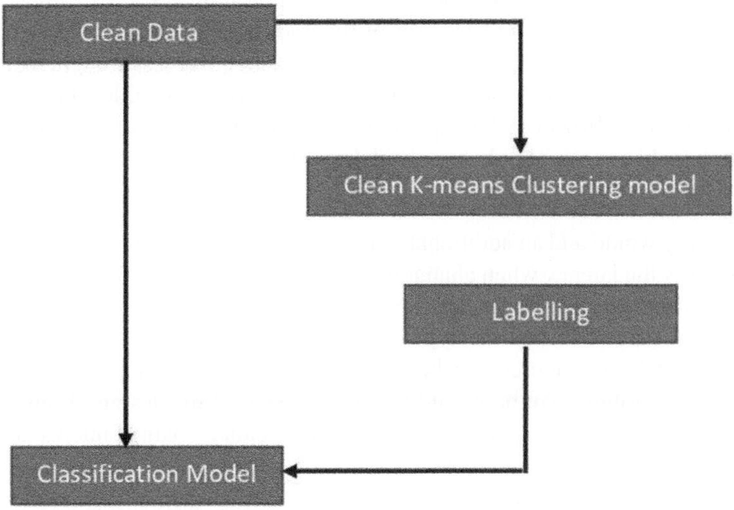

Figure 10.3. *Labeling dataset using K-means clustering algorithm. For a color version of this figure, see www.iste.co.uk/ali-yahiya/tactile.zip*

10.5. Open issues

The TI can be considered as a revolution or an evolution of the type of communications, applications, services as well as business models. It is expected that the TI will be an economy booster for different use cases and scenarios. However, this evolution is still in progress and will need a mature framework that can include all the functionalities described in the previous sections in order to guarantee the low latency which is a key feature of haptic communications. There are still challenges to be addressed in this regard; even this chapter is not an extensive review that gathers all functionalities, but it does provide an idea about the progress in research in this matter, which is a good start towards further elaboration in each mechanism described for reducing the latency. To this aim, there follows some open issues that can be investigated further to achieve this goal:

– Proactive resource allocation would be the best solution for critical applications. The tactile applications can be classified according to their requirements into low, high and adaptive delay applications. The resource allocation mechanism should change its behavior according to the type of application with a dedicated mechanism suitable for each one. Generally, the proactive end-to-end resource allocation stringent latency requirement is the best way to guarantee the 1-ms latency criteria. Mechanisms for scheduling, QoS provisioning, admission control and so on should be re-designed to react as fast as possible to process haptic traffic.

– Mobile core-based cloud native is an interesting idea to bring all the functionalities, capabilities and services of the traditional cloud to the core network. This would reduce lots of hops that the haptic traffic should cross through. In addition, this could bring the cloud closer to the users in terms of physical location and could reduce the latency. However, having an overloaded core network would bring some congestion problems that should be solved through SDN and the orchestration service provided by the VNF.

– Mobility would add an additional dimension of complexity that impacts the QoS and especially the latency when changing the point of attachment from one to another especially when the case of vehicles over TI is considered. The handover delay would influence the type influence the type of communication latency. A position-based solution could be considered the best solution to this type of problem. Artificial intelligence algorithms can be considered at each step of the design of any resource allocation, or orchestration in the core network. The main domain of interest that could be used in combination is enforcement learning or the process of learning, since haptic behavior is repetitive and does not need to generate new action to deal with the motion; therefore, solutions based on learning would be an interesting area to investigate.

– Anticipatory network is also an option to decrease the latency by using principles like context, prediction and optimization. Using the anticipatory network, the context information will then be studied through past and present information, a prediction of the behavior will be achieved, then an optimization is carried out to meet the requirement of any application. This would be useful in the case of the TI especially improving the QoE of users who are involved in haptic communications (Bui *et al.* 2017).

– Artificial intelligence will unquestionably carry a paradigm shift regarding data-oriented approaches; there are still open problems to be resolved. There is no realistic deployment in the TI as there is no comparison for the methods used in learning for each case study. Everything depends on the data set generated by each method and how it is dealt with.

10.6. Conclusion

In addition to the latency issue that we discussed in section 10.2, there are other challenges that face the TI, such as how to keep the TI system as reliable as possible, enabling the system to be scalable, as well as ensuring its security. Indeed, reliability and security are both so important to the system that it should be 99.99% secure when working. With regards to the security, the TI architecture cannot be secured with the traditional techniques of Internet technologies. For example, Tactile Internet actors are vulnerable and not secure against distributed denial of service (DDoS) attacks (which decrease availability), remote hijacking, cloning attacks and man in the middle. Any single TI actor could represent a single point of failure (SPOF) for the entire network and thus damage the availability of data, confidentiality and integrity (Li *et al.* 2018; Mohanta *et al.* 2019).

Security, reliability and availability are such important issues in the TI, if one of them is hijacked, the entire system will be unstable. However, the traditional methods in securing the information system are not convenient for the TI due to it is stringent requirements of QoS, especially for mission critical applications. It is important to re-design methods for ensuring confidentiality, integrity and authentication to be adapted to the TI.

Hidar *et al.* (2021) conceived a model using blockchain for authentication in order to secure human-to-machine interactions, like remote surgery, in the Tactile Internet environment. Thus, a surgeon can now authenticate a robot arm using a good-shared session key and build a high level of security in communication.

10.7. References

Ahmad, H., Islam, M.Z., Haider, A., Ali, R., Kim, H.S. (2021). Intelligent stretch reduction in information-centric networking towards 5G-Tactile Internet realization. eprint: 2103.08856.

Aijaz, A. (2016). Towards 5G-enabled Tactile Internet: Radio resource allocation for haptic communications. *2016 IEEE Wireless Communications and Networking Conference Workshops (WCNCW)*, 145–150.

Aijaz, A., Dawy, Z., Pappas, N., Simsek, M., Oteafy, S., Holland, O. (2018). Toward a Tactile Internet reference architecture: Vision and progress of the IEEE P1918.1 standard. *CoRR*, abs/1807.11915 [Online]. Available at: http://arxiv.org/abs/1807.11915.

Amaral, P., Dinis, J., Pinto, P., Bernardo, L., Tavares, J., Mamede, H.S. (2016). Machine learning in software defined networks: Data collection and traffic classification. *2016 IEEE 24th International Conference on Network Protocols (ICNP)*, 1–5.

Ateya, A.A., Vybornova, A., Kirichek, R., Koucheryavy, A. (2017). Multilevel cloud based Tactile Internet system. *2017 19th International Conference on Advanced Communication Technology (ICACT)*, 105–110.

Bhardwaj, A., Dabeer, O., Chaudhuri, S. (2013). Can we improve over Weber sampling of haptic signals? *2013 Information Theory and Applications Workshop (ITA)*, 1–6.

Bui, N., Cesana, M., Hosseini, S.A., Liao, Q., Malanchini, I., Widmer, J. (2017). A survey of anticipatory mobile networking: Context-based classification, prediction methodologies, and optimization techniques. *IEEE Communications Surveys Tutorials*, 19(3), 1790–1821.

Cabrera, J.A., Schmoll, R., Nguyen, G.T., Pandi, S., Fitzek, F.H.P. (2019). Softwarization and network coding in the mobile edge cloud for the Tactile Internet. *Proceedings of the IEEE*, 107(2), 350–363.

Cen, Z., Mutka, M.W., Zhu, D., Xi, N. (2005). Supermedia transport for teleoperations over overlay networks. In *NETWORKING 2005. Networking Technologies, Services, and Protocols; Performance of Computer and Communication Networks; Mobile and Wireless Communications Systems*, Boutaba, R., Almeroth, K., Puigjaner, R., Shen, S., Black, J.P. (eds). Springer-Verlag, Berlin, Heidelberg.

Cheng, K., Teng, Y., Sun, W., Liu, A., Wang, X. (2018). Energy-efficient joint offloading and wireless resource allocation strategy in multi-MEC server systems. *2018 IEEE International Conference on Communications (ICC)*, 1–6.

Clarke, S., Schillhuber, G., Zaeh, M.F., Ulbrich, H. (2006). Telepresence across delayed networks: A combined prediction and compression approach. *2006 IEEE International Workshop on Haptic Audio Visual Environments and their Applications (HAVE 2006)*, 171–175.

Dab, B., Aitsaadi, N., Langar, R. (2019). Joint optimization of offloading and resource allocation scheme for mobile edge computing. *2019 IEEE Wireless Communications and Networking Conference (WCNC)*, 1–7.

Eid, M., Cha, J., El Saddik, A. (2011). Admux: An adaptive multiplexer for haptic audio visual data communication. *IEEE Transactions on Instrumentation and Measurement*, 60(1), 21–31.

Gholipoor, N., Saeedi, H., Mokari, N. (2018). Cross-layer resource allocation for mixed tactile Internet and traditional data in SCMA based wireless networks. *2018 IEEE Wireless Communications and Networking Conference Workshops (WCNCW)*, 356–361.

Gokhale, V., Dabeer, O., Chaudhuri, S. (2013). HoIP: Haptics over Internet Protocol. *2013 IEEE International Symposium on Haptic Audio Visual Environments and Games (HAVE)*, 45–50.

Gokhale, V., Chaudhuri, S., Dabeer, O. (2015). HoIP: A point-to-point haptic data communication protocol and its evaluation. *2015 Twenty First National Conference on Communications (NCC)*, 1–6.

Gokhale, V., Nair, J., Chaudhuri, S. (2017a). Congestion control for network-aware telehaptic communication. *ACM Transactions on Multimedia Computing, Communications, and Applications*, 13(2), Article 17 [Online]. Available at: https://doi.org/10.1145/3052821.

Gokhale, V., Nair, J., Chaudhuri, S. (2017b). Teleoperation over a shared network: When does it work? *2017 IEEE International Symposium on Haptic, Audio and Visual Environments and Games (HAVE)*, 1–6.

Grasso, C., Karthik Eswar, K.N., Nagaradjane, P., Ramesh, M., Schembra, G. (2021). Designing the tactile support engine to assist time-critical applications at the edge of a 5G network. *Computer Communications*, 166, 226–233 [Online]. Available at: https://www.sciencedirect.com/science/article/pii/S0140366420320077.

Hidar, T., Kalam, A.A.E., Benhadou, S., Mounnan, O. (2021). Using blockchain based authentication solution for the remote surgery in tactile internet. *International Journal of Advanced Computer Science and Applications*, 12(2) [Online]. Available at: http://dx.doi.org/10.14569/IJACSA.2021.0120235.

Holland, O., Steinbach, E., Prasad, R.V., Liu, Q., Dawy, Z., Aijaz, A., Pappas, N., Chandra, K., Rao, V.S., Oteafy, S., Eid, M., Luden, M., Bhardwaj, A., Liu, X., Sachs, J., Araújo, J. (2019). The IEEE 1918.1 tactile internet standards working group and its standards. *Proceedings of the IEEE*, 107(2), 256–279.

Hou, Z., She, C., Li, Y., Quek, T.Q.S., Vucetic, B. (2018). Burstiness aware bandwidth reservation for uplink transmission in tactile internet. *2018 IEEE International Conference on Communications Workshops (ICC Workshops)*, 1–6.

Khemiri, S.K. (2015). 4G traffic offloading through wireless network based on next generation hotspots (NGH). *2015 International Symposium on Networks, Computers and Communications (ISNCC)*, 1–6.

Li, D., Peng, W., Deng, W., Gai, F. (2018). A blockchain-based authentication and security mechanism for IoT. *2018 27th International Conference on Computer Communication and Networks (ICCCN)*, 1–6.

Mohanta, B.K., Jena, D., Panda, S.S., Sobhanayak, S. (2019). Blockchain technology: A survey on applications and security privacy challenges. *Internet of Things*, 8, 100107 [Online]. Available at: https://www.sciencedirect.com/science/article/pii/S2542660518300702.

Monnet, W. and Yahiya, T.A. (2020). HoIP performance for Tactile Internet over 5G networks: A teleoperation case study. *2020 11th International Conference on Network of the Future (NoF)*, 48–54.

Nasir, Q. and Khalil, E. (2012). Perception based adaptive haptic communication protocol (PAHCP). *2012 International Conference on Computer Systems and Industrial Informatics*, 1–6.

Osman, H.A., Eid, M., Iglesias, R., Saddik, A.E. (2007). Alphan: Application layer protocol for haptic networking. *2007 IEEE International Workshop on Haptic, Audio and Visual Environments and Games*, 96–101.

Pérez, G.O., Ebrahimzadeh, A., Maier, M., Hernández, J.A., López, D.L., Veiga, M.F. (2020). Decentralized coordination of converged tactile internet and MEC services in H-CRAN fiber wireless networks. *Journal of Lightwave Technology*, 38(18), 4935–4947.

Rimal, B.P., Van, D.P., Maier, M. (2017). Mobile edge computing empowered fiber-wireless access networks in the 5G era. *IEEE Communications Magazine*, 55(2), 192–200.

Ruan, L. and Wong, E. (2018). Machine intelligence in allocating bandwidth to achieve low-latency performance. *2018 International Conference on Optical Network Design and Modeling (ONDM)*, 226–229.

Sachs, J., Andersson, L.A.A., Araúo, J., Curescu, C., Lundsö, J., Rune, G., Steinbach, E., Wikström, G. (2019). Adaptive 5G low-latency communication for Tactile Internet services. *Proceedings of the IEEE*, 107(2), 325–349.

Saddik, A.E., Orozco, M., Eid, M., Cha, J. (2011). *Haptics Technologies: Bringing Touch to Multimedia*. Springer-Verlag, Berlin, Heidelberg.

Sakr, N., Georganas, N.D., Zhao, J. (2011). Human perception-based data reduction for haptic communication in six-DoF telepresence systems. *IEEE Transactions on Instrumentation and Measurement*, 60(11), 3534–3546.

Sharma, S.K., Woungang, I., Anpalagan, A., Chatzinotas, S. (2020). Toward tactile internet in beyond 5G era: Recent advances, current issues, and future directions. *IEEE Access*, 8, 56948–56991.

She, C. and Yang, C. (2016). Ensuring the quality-of-service of tactile internet. *2016 IEEE 83rd Vehicular Technology Conference (VTC Spring)*, 1–5.

She, C., Yang, C., Quek, T.Q.S. (2016a). Cross-layer transmission design for tactile internet. *2016 IEEE Global Communications Conference (GLOBECOM)*, 1–6.

She, C., Yang, C., Quek, T.Q.S. (2016b). Uplink transmission design with massive machine type devices in tactile internet. *CoRR*, abs/1610.02816.

Sheng, Y. (2018). Scalable intelligence-enabled networking with traffic engineering in 5G scenarios for future audio-visual-tactile internet. *IEEE Access*, 6, 30378–30391.

Swamy, V.N., Rigge, P., Ranade, G., Sahai, A., Nikolic, B. (2016). Network coding for high-reliability low-latency wireless control. *2016 IEEE Wireless Communications and Networking Conference*, 1–7.

Szabo, D., Gulyas, A., Fitzek, F.H.P., Lucani, D.E. (2015). Towards the tactile internet: Decreasing communication latency with network coding and software defined networking. *Proceedings of European Wireless 2015; 21th European Wireless Conference*, 1–6.

Wirz, R., Ferre, M., Marín, R., Barrio, J., Claver, J.M., Ortego, J. (2008). Efficient transport protocol for networked haptics applications. In *Haptics: Perception, Devices and Scenarios*, Ferre, M. (ed.). Springer-Verlag, Berlin, Heidelberg.

Yu, Z., Gong, Y., Gong, S., Guo, Y. (2020). Joint task offloading and resource allocation in UAV-enabled mobile edge computing. *IEEE Internet of Things Journal*, 7(4), 3147–3159.

Zhao, Z., Zhao, R., Xia, J., Lei, X., Li, D., Yuen, C., Fan, L. (2020). A novel framework of three-hierarchical offloading optimization for MEC in industrial IoT networks. *IEEE Transactions on Industrial Informatics*, 16(8), 5424–5434.

List of Authors

Ian F. AKYILDIZ
School of Electrical and Computer Engineering
Georgia Institute of Technology
Atlanta
USA

Tara ALI-YAHIYA
Department of Computer Science
University of Paris-Saclay
France

Bakhtiar M. AMIN
Department of Computer Science and Engineering
University of Kurdistan Hewlêr
Erbil
Iraq

Pinar KIRCI
Department of Computer Engineering
Bursa Uludağ University
Turkey

Wrya MONNET
Department of Computer Science and Engineering
University of Kurdistan Hewlêr
Erbil
Iraq

Nicola Roberto ZEMA
Department of Computer Science
University of Paris-Saclay
France

Index

K

L

M

N

O

P

Q

R

S

T

U

V

W, Z

Printed and bound by CPI Group (UK) Ltd, Croydon, CR0 4YY
27/10/2024
14580317-0001